W9-ABQ-731

The North American Porcupine

SECOND EDITION

The North American Porcupine

SECOND EDITION

Uldis Roze

Comstock Publishing Associates
a division of
Cornell University Press
Ithaca and London

Published 2009 by Cornell University Press

Printed in the United States of America

Library of Congress Cataloging-in-Publication Data

Roze, Uldis, 1938–
The North American porcupine / Uldis Roze. — 2nd ed.
p. cm.
Includes bibliographical references and index.
ISBN 978-0-8014-4646-7 (cloth : alk. paper)
1. North American porcupine. I. Title.
QL737.R652R69 2009
599.35'974—dc22
2008055510

Cornell University Press strives to use environmentally responsible suppliers and materials to the fullest extent possible in the publishing of its books. Such materials include vegetable-based, low-VOC inks and acid-free papers that are recycled, totally chlorine-free, or partly composed of nonwood fibers. For further information, visit our website at www.cornellpress.cornell.edu.

Cloth printing 10 9 8 7 6 5 4 3 2 1

Contents

Preface to the Second Edition

A preface is usually the last part of a book to be written, and it looks backward and summarizes the project. In the spirit of looking back, I recall that when I was a young graduate student working for a degree in biochemistry, the great biological question I hoped to take part in solving was the nature of the genetic code: how did the DNA code of nucleotide sequence translate into the protein code of amino acid sequence? The question was solved in 1961, midway through my graduate study, by Heinrich Matthaei and Marshall Nirenberg at the National Institutes of Health (NIH). What was unexpected about their discovery was how the problem was solved—not by brute-force sequencing of nucleic acids and proteins but by the use of a cell-free system that translated an artificial RNA, polyuridylic acid, into an artificial protein, polyphenylalanine. The rest of the code was cracked in a few months.

My own postgraduate scientific wanderings led me not on to the biology of the gene but into the biology of forests and their porcupines. It has been a wonderful journey, as full of surprise and delight as the better-known adventure of molecular biology. Let me offer three examples, all uncovered since the first edition and described more fully in subsequent chapters of this book.

Porcupines occupy some of the northernmost species range in North America, up to the tree line in Canada and beyond it in regions of Alaska. To survive the punishing winters, porcupines seek shelter in winter dens. But some individuals do not use dens, spending the winter in evergreen cover. Are these Rambo porcupines, able to survive greater punishment than their den-using conspecifics?

To compare the environment of the den with that of evergreen cover, Dominique Berteaux and coworkers in Quebec constructed two artificial porcupines, kept internally warm by electric heat and insulated from the cold by the pelts of two animals that had died naturally the year before. The artificial porcupines, whose power demand could be externally monitored, demonstrated that, at very low winter temperatures, the two environments were operationally similar (Chapter 6).

Another problem faced by porcupines is surviving on a diet of tree leaves and bark. Trees do not wait for animals to eat them—they fight back by infusing their tissues with a bitter mix of poisons. How does a porcupine survive this challenge? It does so by adding some clay to its normal diet. Clay detoxifies plant tissues by binding with their tannins and alkaloids (Chapter 13).

A third problem faced by porcupines is assembling a normal intestinal flora that is appropriate to their dietary needs. The normal flora is the complex of intestinal bacteria that porcupines depend on for digestion of the complex polymers (including cellulose) that are present in their plant diets; differing diets require differing normal floras. A porcupine whom I was accompanying in the woods accomplished this in an elegant way: it ingested a piece of a porcupine dropping left by a previous inhabitant of the den my porcupine was going to settle in. Porcupine droppings are dry pellets composed of undigested fiber and the intestinal normal flora. By inoculating itself with the previous inhabitant's gut bacteria, my porcupine was adapting itself to the local food supply (Chapter 13).

A striking impression of my later studies of the porcupine has been something I was only beginning to appreciate in the 1980s—the length and complexity of their lives (Chapter 14). This has reinforced my decision to rule out all experimentation that would cause harm to my animals. I have not regretted this decision.

Many people have supported both the research base and the editorial work involved in this second edition. Foremost among these have been my two long-time collaborators, David Locke of Queens College and David Chapman of Lakehead University. They have contributed generously their ideas, expertise, encouragement, and many hours in their labs. Porcupine biology would be poorer without their contributions. Other people who have contributed to the present work include Bruno Anthony, Barbara Bellens-Picon, Dominique Berteaux, Ann Bryant, George Burton, Marie Byrne, John Dennehy, Susan Giglia, Andrew Greller, Bernd Heinrich, Kathleen LaMattina, Stuart Landry, Kam Leung, Guang Li, Heidi Lovette, Wayne Lynch, PoKay Ma, Geraldine Mabille, Dimitri Matassov, Leslie Marcus, Craig Packer, Kate Pechenkina, Cesar Sanchez, Evelyn Silverman, Rick Sweitzer, Phil Vollono, and Heidi Zapata. In addition, I acknowledge the long-term funding of the Professional Staff Congress–City University of New York (PSC-CUNY) Fund. My overarching thanks goes to my wife, Steph, who has shared hikes and a passion for the porcupine and to whom this book is dedicated.

Preface to the First Edition

A study of the porcupine requires, above everything, time—not only time to follow a population through an annual cycle of energy and mineral acquisition, reproduction, and winter denning but also much longer blocks of time to work out life-history strategies in these long-lived animals: reproductive success, territorial expansion or contraction, dispersal. Those kinds of data are needed to understand social structure and ecological strategies. I have had the luxury of such time, in the form of two sabbatical leaves from Queens College and a liberal teaching schedule that allowed weekly visits to the woods. The study here described began during my sabbatical of 1978–1979, the writing of this book during my sabbatical of 1985–1986.

A porcupine study also requires a reliable access to porcupines. My wife and I laid the groundwork for that in 1970, with our purchase of 70 acres of forested Catskills mountainside. We built our own cabin, dug a well, cut firewood. Above us stretched miles of unbroken forest, without human dwellings and with vistas from the mountaintops such as the Indians must have seen. After the hunting season, winter snows showed no human tracks other than my own. But there were other inhabitants in the woods. A black bear behind the house turned over 150-pound rocks searching for grubs. A ruby-throated hummingbird came to feed on trumpet creeper. And one October morning, I discovered a small porcupine in the leafless crown of an apple tree.

My porcupine encounters were rare events in the early years because I was not seeking out porcupines; the animals, at times, were seeking out me. From 1978, I learned how to find porcupines on my own by locating them in dens in the winter, at salt sources in the spring, and in apple orchards in the summer. With those techniques alone, I could encounter upwards of 7 animals a day. However, I was not yet marking individual animals. A grant from Queens College in 1981 and subsequent grants from PSC—City University of New York allowed me to radiocollar animals beginning in 1981. I marked others with eartags and tattoos. I now discovered my study population was well in excess of 7 animals; in 1988, I encountered 39 different animals. An artificial salthouse, as described in Chapter 5, eased my access to animals. Previously, I had sought porcupines at three barns, four assorted outbuildings, and our own cabin on occasion. After building the salthouse, I could conveniently visit a single location.

For capture, I discovered the efficiency of rubberized gloves and a 12-quart plastic ice chest with closable top. I baited my live traps with Red Delicious apples and used aluminum flashing to guide animals into tree traps. My commercial radiocollars were mounted on soft nylon braid and hung around the necks of the animals. From Wendell Dodge, I learned to use ketamine for safe field anesthesia. From Aelita Pinter came the oxytocin technique for establishing lactation status. From Hans Kruuk, I borrowed the technique of permanent identification by groin tattoo.

In addition, many people have helped with every aspect of the study, from discussion and ideas to assistance in the field or laboratory: Sheldon Aaronson, David Alsop, Sydney Anderson, Pat Bridges, Helen Cairns, Robert Calhoon, Peter Chabora, Tom Chen, Michael Crooks, Edward Dolensek, Jean Dorst, Robert C. Dowler, Durland Fish, Wayne Graff, Andrew Greller, Max Hecht, Lyndon Hill, Elizabeth Horner, Walter Huggans, Frank A. Iwen, Stuart Landry, Rene Lavocat, Richard Lindroth, David C. Locke, Leslie Marcus, George Mason, Cathy McConnell, Daniel McKinley, Martha Mixon, Milton Nathanson, Dick Pardi, Michael Seidel, Ward B. Stone, George Szilagyi, Donald J. Tipper, Nick Vatakis, and Pam and Judd Weisberg. Wendell Dodge, Roger Powell, and Louise Emmons critically reviewed drafts of Chapters 9, 2 and 7, and 10 and 12, respectively. My brother Maris read the entire manuscript, as did two professional mammalogists. John Winsch supplied the drawings and figures. Finally, I thank my wife, Stephanie, for the unfailing support and stimulating atmosphere that made this study possible.

The book that follows has been a pleasure to write. I hope the reader will find it an equal pleasure to read. But many more books on the porcupine remain to be written because again and again our understanding of these animals, like our understanding of much of the world we live in, ends in a mystery.

The North American Porcupine SECOND EDITION

1 Porcupines and People

It is a warm summer night, and I am about to meet my first porcupines. My wife and I are sleeping in the cabin we have just finished on the edge of the woods stretching up a Catskills mountainside. Sometime after midnight, an insistent sound punches through my sleep. I rise up on one elbow and listen. The sound comes again—a deep, resonant rasping from the far corner of the cabin.

I put on my glasses, pick up a flashlight, and go out in the night. The cone of my light reveals two large animals reaching over the foundation and leaning against the wall of the cabin. My rational mind tells me they are porcupines, but they look like creatures out of a dream. They stare back at me with the ragged outlines of the unbelievable, animals more surreal than wild.

Then the rage of the property owner is upon me. It's our new house they are chewing! I pick up a stick, raise it high, and bring it down on the nearest animal. But my aggressiveness leaves me as my glasses fall off in the dark, and I suddenly realize I am both naked and blind. The porcupines escape. Only a scatter of quills on the ground and a patch of damaged plywood on the cabin next morning tell me I was not dreaming.

The porcupines returned, and I am ashamed to say I clubbed several before working out a simple solution—stapling a 2-foot belt of chicken wire all around the base of the cabin. The porcupines had been interested only in the plywood walls. With the bait covered, they stopped coming.

My attitudes of the 1970s, so embarrassing to me at present, were probably typical among people in the region. Even neighbors who do not believe in hunting deer have told me they kill porcupines because the animals damage trucks, outbuildings, and property in general. In other words, the animals are considered pests.

Native American–Porcupine Relationships

The porcupine has not always been considered a pest. To the Indian tribes that inhabited North America before the European settlement, the porcupine was a precious resource, treated with reverence by some. Among the woodland tribes, the porcupine held an important place in food-gathering and the economy. That was especially true of the Micmacs of New Brunswick and Nova Scotia and of the Montagnais of eastern Quebec, who were collectively known as the Porcupine Indians (Speck and Dexter 1951). These northern woodland Indians subsisted largely on fish, mollusks, and wild plants in the summer and on big game such as moose and woodland caribou in the winter. The big game was hunted by driving the animals in deep snow (Thwaites 1907; Speck and Dexter 1951). Although they gathered wild roots, fruits, and nuts as available, the Indians did not have an agricultural economy, unlike the Hurons and the Iroquois to the south. The vulnerable period in their hunting-fishing-gathering economy was early winter, when ice restricted fishing but adequate snow cover had not yet accumulated for driving moose and caribou. Then the porcupine became a critical link in the Indians' survival. Along with the beaver, it remained the only animal that could be hunted with relative efficiency.

A fortunate historic accident has left a well-drawn picture of the original lifestyles of the woodland Indians. Almost from the beginnings of the French settlement in present-day Canada, the settlers were accompanied by Jesuit missionaries. Beginning in 1610, these articulate and highly educated men sailed from France to live among the Indians. Although primarily concerned with their Christianizing mission, they recorded in rich detail the techniques of hunting and gathering, social customs, religious beliefs, warfare, commercial relationships, crafts, and much else. Here Father LeJeune, writing in 1633 from Quebec, describes winter hunting of the porcupine:

> The porcupine is taken in a trap, or by coursing. The dog having discovered it, [the porcupine] is sure to be killed if it is not very near its abode, which it makes under large rocks; having reached this, it is in a place of safety, for neither men nor dogs can crawl into it. It cannot run upon the snow, and is therefore very soon put to death.
>
> They singe them as we do pigs in France, and after they are scraped, they are boiled or roasted, and are quite edible, although rather tough, especially the old ones, but the young ones are tender and delicate. But in taste they are not equal to either our wild boar or our common pig. (Thwaites 1907)

Every few years the game failed, the Indians starved, and the Jesuits recorded the painful extremities to which they were pushed: the Indians would eat their valuable dogs, their dressed elkskins, and finally their shoes. The weak (usually the children) would begin to die. At such times of famine, the porcupine acquired an almost supernatural significance. Again, Father LeJeune recounts an experience. He and his party of Indians had entered the vast forests of Quebec in November 1633 and had been forced to move camp repeatedly because the thin snow made big-game hunting impossible. The party was starving.

> On the 24th of December, the evening before the birth of our Savior, we broke up [camp] for the seventh time. We departed without eating, and journeyed for a long, long time, then worked at house-building; and for our supper Our Lord gave us a Porcupine as large as a suckling pig, and a hare. It was not much for 18 or 20 people, it is true, but the holy Virgin and her glorious Spouse, saint Joseph, were not so well-treated on the same day in the stable at Bethlehem. (Thwaites 1907)

Because on many occasions the porcupine saved the Indians' lives, they believed that the animal offered itself to the hunter and must be treated with gratitude and respect. The bones of beavers (whose ecological relationship with the Indians was similar) and porcupines could not be given to the dogs but had to be burned or thrown into the river. Otherwise, the animals would make themselves hard to capture in the future. A residue of that belief is retained in the woodlands of Quebec, northern Michigan, and the Maritimes today. Many residents believe that the unprovoked killing of a porcupine is a crime because a porcupine is almost the only animal an unarmed man lost in the woods can capture for food.

Beyond its life-sustaining function, the porcupine played a second important role in the lives of the North American Indians. Its quills, both in the loose state or as elaborate quillwork, came closest to serving as a medium of exchange. Such currency was supplanted during the settlement period by wampum, a European introduction. Porcupine quillwork is an exacting, time-consuming craft. The quills were collected, sorted according to size, dyed with vegetable dyes, and then woven into a ground of deerskin or birch bark. They were held in place by split spruce root or sinew binding, or by forcing them through tiny holes punched in the ground material with a bone awl. The finished article presented a flat, durable, close-grained surface that breathed with subtle natural color (Orchard 1971; Odle 1971). The dyes used for porcupine quills came from a wide range of native trees and wildflowers. Ash and earths were often added as mordants. Table 1.1 lists the sources exploited by the Ojibwa of the Great Lakes region. The dyeing process typically required long periods of boiling of quills in the plant extracts. If the color obtained was too pale, the treatment was repeated.

Table 1.1. Ojibwa quill dyes and mordants

Plant	Part used	Color
Alder (*Alnus incana*)	Inner bark used with bloodroot, wild plum, red osier dogwood	Red, yellow
Bloodroot (*Sanguinaria canadensis*)	Root	Red
Bur oak (*Quercus macrocarpa*)	Inner part with hazel burrs, butternut inner bark	Black
Red osier dogwood (*Cornus stolonifera*)	Inner bark with ashes of birch, oak, cedar bark	Red
Goldthread (*Coptis groenlandica*)	Root	Yellow
Lichens (*Usnea barbata*)	Whole plant	Yellow
Maple (*Acer* spp.)	Rotted wood	Purple
Puccoon (*Lithospermum carolinense*)	Dried root with ocher	Red
Sumac (*Rhus glabra*)	Inner bark, pulp of stalk	Yellow
White birch (*Betula papyrifera*)	Inner bark with ashes of dogwood, oak, and cedar bark	Red
American plum (*Prunus americana*)	Inner bark with bloodroot root, red osier dogwood bark, alder bark	Bright red

Source: Lyford (1943)

Quillwork found its way into belts, leggings, blouses, boxes and trunks, place mats, bracelets, necklaces, moccasins, purses, medicine bags, and tobacco bags. An example of the intricacy and vibrant color possible in quillwork is shown in the Sioux arrow case in Color Plate 1. Although most of the articles were small, the amount of work required was staggering. The finest quillwork was constructed of small, closely spaced quills that modulated complex patterns in the natural pigments of the plant world. Such work often has an unpredictability that is more startling for being set in a convoluted pattern. Perhaps the worker ran out of green-dyed quills and made a small substitution of yellow, or perhaps, like Navaho weavers, the quillworker broke the patterns as a philosophical statement. Such work then takes on the qualities of music.

References to porcupine quillwork often find their way into Indian myths and fairy tales. Thus, a Cheyenne myth explains the origin of the Big Dipper as follows. A beautiful young girl, growing up as an only child, wanted brothers. She fashioned seven buckskin outfits with splendid quillwork and presented them to seven brothers who lived alone in distant woods. Delighted with the gift, they accepted her as a sister. But a monstrous buffalo heard about the girl, grew jealous, and set out to capture her for himself. Too weak to resist, the girl and her seven brothers used magic to escape into the sky, where they shine in their quillwork outfits as the eight stars of the Big Dipper, the sister being the brightest (Erdoes and Ortiz 1984).

Porcupine quillwork extended well into areas where the porcupine may have been absent or marginal. Tribes such as the Sioux of the Dakotas, Winnebago of Nebraska, Sauk of Iowa, and Potawatomi of Kansas all inhabited territories where the woodland habitat needed by porcupines was scarce or absent. Yet all contained active centers of porcupine quillwork. This required trade with tribes that coexisted with the porcupine: the Micmac and Huron in the Northeast, the Ojibwa and Menominee in the Midwest, the Shoshoni and Nez Percé in the West (Orchard 1971). Thus the porcupine, in centuries before the European settlement, served a cohesive function on the North American continent. Its quills formed the first currency, valuable because quillwork was beautiful and hard to make.

Indians found the porcupine useful in smaller ways as well. One example is the porcupine combs that the Crow Indians manufactured from the underside of the tail, which is covered not by quills but by firmly anchored, stout bristles used by the animals to climb trees. George P. Belden, in an 1872 account, describes the combs: "They are very simple, and consist of a Hedgehog's (i.e., porcupine's) tail, the bristles serving as teeth. When a Hog is killed, the tail is skinned off the bone, and a wooden handle is inserted. When dry, it is ready to use, and is by no means a bad substitute for the bone or horn combs we use. A Hedgehog comb is an indispensable article to every Indian girl, as it enables her to keep her long black hair in order" (quoted in Seton 1928, 629). Another porcupine application involved the Indian craft of maple sugaring. The Indians, who lacked metal spiles for drawing off the maple sap, used hollowed-out porcupine quills instead (Denys 1908; Speck and Dexter 1951). Clearly, in ways both large and small, the porcupine was a resource to the Indians and therefore an animal to be respected.

Contemporary Porcupine–Human Interactions

With the disappearance of the cultures and economies of the Indians, the status of the porcupine fell. In no state in the United States today is the animal protected from hunting at any time. In the past, many states offered bounties for porcupines killed, and some undertook massive poisoning programs. The justification for eradication campaigns has been economic. There are two main areas of conflict. The first involves the extensive modern use of high-sodium materials, especially plywood in construction and salt for de-icing roads.

Unlike the low-sodium diets available for humans, no low-sodium plywood is available for porcupine-inhabited areas, although such a plywood would be easy to manufacture. Road salt is used on an increasing scale to control winter ice. The salt accumulates on the underbodies of cars and trucks, including tires, hoses, and electrical wiring. In addition, synthetic rubber contains endogenous sodium as part of the manufacturing process. When porcupines find these salt morsels in the spring, they can disable a vehicle after a night's chewing.

The porcupine's location of sodium sources is unerring. A neighbor, who runs a camp for young people, observed with some puzzlement that porcupines attacked the wooden outer walls of boys' but not girls' dormitories. I asked him, "Which sex is most likely to urinate against the wall of the dormitory at night?"

The second major human–porcupine conflict involves the timber industry. Porcupines feed on the bark of trees in winter. Most of their feeding takes place in the crown of the tree, where they may girdle individual branches or the leader of the tree but not the tree itself. The tree that survives many years of porcupine attacks takes on an eerie, witchlike quality (see Figure 1.1). The trunk is shortened, and branches twist in baroque unpredictability. Although often of short stature, the trees may be quite old, like some of the pollarded trees of the European countryside. Such "witch trees" are found in local clusters in the vicinity of porcupine dens and are best seen during their leafless state in winter. From the porcupine's perspective, a grove of witch trees represents good forest management. The short trees are easy to climb, the thick sprouting of young branches offers food that is easily harvested, and the low, thick crowns provide some shelter from winter wind.

Forest well managed from the porcupine's point of view may not be well managed from the lumberman's point of view. Low trees with twisted crowns produce reduced lumber yields, especially of the long, knot-free timber that lumbermen prefer. For that reason, the porcupine is persecuted wherever it occurs, even when it feeds on trees such as beech and hemlock, which have minimal commercial importance. In many New England states, bounties were formerly offered. Vermont paid out $90,000 in 1952 alone, the last year it had a bounty law. That represents many animals killed at 50 cents per snout. Most bounty laws were repealed during the 1920s, although New Hampshire, the last state with a bounty law, ended the practice only in 1979 (Dodge 1982).

Figure 1.1. Porcupine in linden tree shaped by past feeding. (Photo by Uldis Roze [hereafter UR])

The bounty system suffered from the energies of private enterprise. William A. Reeks (1942) of New Brunswick mentions problems such as animals brought in from outside the prescribed bounty area and entrepreneurs who have been able to manufacture porcupine snouts out of forepaws and thus multiply each bounty several-fold. For three winters beginning in 1957–1958, the state of Vermont embarked on a porcupine poisoning program that led to the deaths of 10,800 animals (including 1700 that were shot). The poison used was a sodium arsenite gel injected into apples. In 3 years, 32,000 poisoned apples were placed in porcupine dens (Faulkner and Dodge 1962). Many states, including California, Idaho, Oregon, Montana, Michigan, Wisconsin, New York, and Vermont, are experimenting with biological control by the introduction of fisher into areas from which the animal had been previously trapped out (Dodge 1982; Conniff 1986). As described more fully in the following chapter, the fisher appears to be exerting significant control in Michigan and elsewhere. In most states, the experiment is proceeding blindly, with no monitoring of the populations involved.

Can these economic conflicts be lessened? Let us consider the conflict over sodium use. Plywood becomes porcupine-proof simply by substituting potassium for sodium as the cation in the resin. The national market for exterior plywood is so large that the porcupine-vulnerable segment may not justify a change for the entire industry. But for a smaller manufacturer, such a product may represent an economic opportunity.

With respect to road salt, the economic arguments are less clear-cut, partly because of the nature of the accounting system used. Road salt is applied by highway departments to clear winter ice. Such an agency has a budget for snow and ice removal and is likely to ask a single question: Which chemical will melt ice at lowest cost? The answer is rock salt—sodium chloride. But the use of rock salt entails hidden costs. Apart from its porcupine connection, rock salt deteriorates concrete, corrodes underground power lines and steel supports, rusts out car bodies, harms street trees, and pollutes water supplies. Some of those costs came due in summer 1988 when New York City had to close the Williamsburg Bridge across East River because of rusted-out steel supports. Engineers blamed rock salt. None of those costs is normally factored into snow-removal budgets. In 1986, rock salt delivered in bulk in New York cost $34 per ton. But if indirect costs are factored in, estimates place the total cost at upward of $1000 a ton. At that price, alternate de-icing techniques become competitive. None of them—sand, urea, calcium, or calcium-magnesium chloride—has the porcupine involvements that sodium chloride does.

Similar economic arguments have ended the state bounty systems and poisoning programs to control timber damage by porcupines. On the one hand, biological control by the introduction of fisher appears to be cheap and effective. On the other hand, heavy damage by porcupines is generally restricted to the vicinity of dens. In mountainous regions, dens are located in steep and inaccessible terrain, where lumbering is uneconomical. It has therefore become obvious to the agencies concerned that the damage caused by porcupines is too small to justify the costs of active control.

Whereas Indians saw in the porcupine a valuable resource, contemporary Americans have treated the animal as a pest that destroys property and damages timber, even though part of that economic damage has been self-inflicted. The two attitudes represent opposite sides of a coin; both are based on the economic benefit or cost of the animal.

Study of the Porcupine

A third kind of relationship is possible with porcupines and with the natural world in which they live, one that I champion in this book—to study them. Ours is a society that encourages inquiry, in part because it can lead to tangible returns. The woods, rivers, mountains, and prairies that surround our cities remain filled with astonishing life stories: insect pupae that spin around inside their cocoons to foil would-be wasp parasitoids, Jack-in-the-pulpit flowers that change sex depending on success during the previous year and that lure fungus gnats to their deaths in female flowers but leave an escape door in males, and carrion beetles that carry a platoon of abdominal mites to win the battle against blowfly larvae invading the same carcass. Because the woods are filled with stories, many still undiscovered, careful observation of almost any organism is likely to yield surprise and delight.

The Catskills land we bought in 1970 was part of an abandoned farm. An aerial photo of the area, taken by the U.S. Department of Agriculture in 1959, clearly shows surrounding roads, trails, farmhouses, even individual trees in orchards. Adjacent to Beech Ridge Road lie the remaining fields and pastures of the Decker farm, with the great woods reclaiming the rest. In the pasture, hay-cutting is in progress; a border of white encircles the darker uncut grass of the center. The camera caught Winnie Decker in one of the last farm chores of his life. He died the following year, and the farm passed into nonfarming hands. But these hillside acres, which supported the Decker family for almost 50 years and the Angle family for 100 years before that, remain productive. The crop they are yielding today, and can continue to yield for centuries to come, is a crop of unfolding knowledge, a crop as nourishing to the human spirit as fresh bread is to the body.

Whether the Indians systematically studied the porcupine we do not know because they left no written records. Their cultures have been largely swept away, and their knowledge with them. That knowledge must have been considerable because it was required for survival. Early European travelers and missionaries did leave a written record. Their observations, although sometimes perceptive, tend to be anecdotal; they are seldom presented in quantitative terms, and there are no studies over the long term. The one notable exception is a 1727 study of the North American porcupine, the first to examine the animal with scientific rigor. The author, Michel Sarrazin, was physician to the king at Quebec, a correspondent of the Royal Academy of Sciences at Paris, and a contemporary of Isaac Newton and R. A. F. de Réaumur (who communicated the article to the academy; de Réaumur 1727).

Dr. Sarrazin's study of the porcupine set a high standard. He observed that the Canada porcupine belongs to the group of animals that gnaws (the rodents) and that it differs in important ways from the African porcupine. It feeds on the bark of living trees, never dead ones, and prefers above all certain pines and the white cedar. The animal also feeds by grazing. Sarrazin dissected the animal and described its internal anatomy in detail. He described its ability to erect its quills, whose tips he observed under a new instrument, the microscope. He discussed the "great question" of whether the porcupine can throw its quills and disproved the idea. He explained why porcupine quills travel in only one direction in the body. He examined the popular belief that porcupines mate in the trees with the female suspended head down from a branch and the male hanging right side up from another branch. He dismissed the theory, but added modestly that he could not describe the true process in detail. He stated that the animal lives for 12 to 15 years. He correctly observed that porcupines have a gestation period of 7 months and bear a single young.

Sarrazin's astonishing paper, inventive in its experimental approach and filled with details of taxonomy, anatomy, and natural history, was correct on every major point it considered. Some of the findings were not confirmed until more than 200 years later. It is therefore unfortunate that Sarrazin's work is seldom remembered today in connection with the porcupine. He is, however, widely commemorated in another context. Named after him are the Sarraceniaceae, the family of carnivorous pitcher plants of sphagnum bogs, whose North American range stretches from Quebec south to Florida and west to the Pacific coast (Rousseau 1957; Gariepy 1961).

Many scientists and naturalists have followed in Sarrazin's footsteps as students of the porcupine. An outstanding figure was Walter Penn Taylor, whose studies were published more than 200 years later, in 1935. His 177-page treatise carries the no-compromise title "Ecology and Life History of the Porcupine (*Erethizon epixanthum*) as Related to the Forests of Arizona and the Southwestern United States" (Taylor 1935). But from the start, Taylor served notice that he intended to pay attention to more than the peeled bark of trees: "In this study, approached from the standpoint of ecology, an attempt was made to record all facts of possible significance, whether or not they appeared to have definite economic bearing" (5). Taylor's concern was not the discovery of ways to control the porcupine. It was, rather, "an attempt to determine the actual role of a woods-inhabiting rodent in the ecology of the forest" (9). All aspects of the animal's life came under scrutiny: feeding preferences, annual migrations, navigational abilities, mating behavior, and techniques of capture in the field (Taylor recommended climbing up after the animal into its tree and tickling its nose with a pine branch to drive it back toward the pursuer). His account remains valuable not only because it is thorough but also because it retains the freshness of the woods and mountains he traveled for so many years.

Large-scale studies of the porcupine that have followed Taylor's include the work of Albert Shadle (1948, 1951; reproductive physiology), James D. Curtis (1941, 1944; porcupine–timber relationships), Robert Brander (1973; Clarke and Brander 1973; environmental physiology), and Wendell Dodge (1967, 1982; ecology and reproduction). More recent comprehensive studies have benefited from the technique of radiotelemetry. In addition to my own, major studies include those of Rick Sweitzer in Nevada, Todd Fuller in Massachusetts, and Dominique Berteaux in Quebec. Each has added fresh insight and shed a bright light on the subject. This book is an attempt to follow in that tradition. Yet the topic is far from exhausted. The scientific revolution known as molecular genetics has hardly touched the porcupine. When it does so, we can expect a further revolution in our understanding of this animal.

Aims of the Present Study

Naturalists have been studying the porcupine for more than 250 years, so the question might be raised whether there is anything left to discover. Indeed there is. The questions a naturalist would like to ask often cannot be answered at the time because the needed technical or conceptual infrastructure is lacking. In natural history, the realm of the possible is continually expanding. Just as the scanning and transmission electron microscopes have revolutionized the study of the cell, so new instruments and new techniques have made a new set of questions possible in natural history.

One of those powerful techniques is radiotelemetry. A small radio transmitter is attached to an animal that is then released in the wild (see Figure 1.2). With a directional antenna and tunable receiver, researchers can locate the animal wherever it wanders. They can then answer questions about food sources, den sites, mating interactions, territoriality, activity budgets, and a mass of other field biology subjects. The study described here represents the longest continuous radiotelemetry study of porcupines reported in the scientific literature. At the time of this writing, the telemetry study has been running for 25 years. One old female was monitored for most of her 21 years of life; her life story is presented in the last chapter of this book. Such long periods of observation require faith that something new will be discovered every year, that lives will not slip into predictable ruts. That faith has indeed been justified.

Two males, each followed for 5 years and showing different degrees of success and failure, suddenly emigrated. By doing so, they forced a revision of my understanding of social structure. Another male was found drowned in a river. An autopsy revealed the cause of death. The flesh around the muzzle and lower jaw was reddened from the impact of a fall, and the skull was cracked in two places where he had hit projecting boulders in the stream. He had fallen from a hemlock tree leaning over the edge of the stream, where I had seen him the previous week. But why did he fall? The head of the left ilium (hip bone) showed an older fracture, half healed. It may have represented a preceding fall and must have left half his body painful and weakened. In confirmation, the claws of the right front paw were heavily worn; claws of the

Figure 1.2. Three-month-old female with radiocollar. (Photo by UR)

left side were overgrown because the left arm was being favored. Hence, the fatal fall had been brought on by injuries suffered in an earlier accident. The idea of tree-climbing animals falling out of trees sent me to the museums to check porcupine skeletons for healed fractures. More than one-third of the specimens showed healed fractures. But if animals fall so often, should they not impale themselves on their own quills? Another surprise—the quills have antibiotic properties.

I remember my surprise and delight when I found my first niptwig under a porcupine feeding tree. A niptwig is a terminal twig, stripped of its leaf blades but with petioles still attached, which the porcupine nips off and discards at the end of its meal (see Figure 1.3). By looking for niptwigs, I could find out not only which trees a porcupine was feeding on but also how intensively the tree was being used. And by realizing how far out on the branches the animals went to feed, I could understand why they risked falls. As the years have gone by, perhaps the most astonishing aspect of this animal has been its ability to astonish—to disprove old assumptions and raise deeper mysteries. More recent studies have shed light on the quill, including the mechanisms for erection, warning-odor production, and detachment. And completely unexpected behaviors have come to light: geophagy, coprophagy, and the role of urine in signaling.

Figure 1.3. Linden niptwigs on the ground. Petioles and portions of leaf midrib are left unconsumed. (Photo by UR)

Partly out of a growing understanding of the complexity of their lives, I have refrained from certain kinds of deliberate experiments—those that kill or injure my porcupines. That has closed some avenues of investigation, but it is my belief that only the living, free animal can raise the most interesting questions because it holds the power of the unexpected and because I meet it in its own world. A study of dead porcupines would not have told me which animals know one another, how they interact, and how they respond to the changing forest.

It is nevertheless fair to say that a sophisticated understanding of any organism has not depended entirely on the study of living individuals. The museum skeletons that confirmed my theories about porcupines' falling out of trees were prepared from dead specimens. The essential digestive role of the caecum was elucidated by a study of sacrificed animals, as was the important role of the accessory corpora lutea in maintaining a long pregnancy, of specially developed muscles used in climbing, of the specialized male accessory sex organs used in hystricomorph reproduction, and so on. Although the list is long, porcupine biology today remains much less well-developed than human biology. And if we have gained our knowledge of the latter without sacrificing humans for experimental purposes, why can we not do the same for porcupines?

A naturalist lives in two worlds: the world of nature and the human world of ideas and shared values. One world is represented by a fallen tree on a mountain ridge, the other by the library, the laboratory, and conversations with fellow scientists. In this study, I have tried to do justice to both. I have also tried hard to listen to the porcupine and to observe it free of the constrictions of my own hypotheses. Francis C. Evans (1985), past president of the Ecological Society of America, has pointed to the critical role of natural history in ecological thought: "Whatever our chosen frame of reference with respect to observation may be, there is always some larger frame that is possible, and there is always a chance that this larger frame may contain something we did not anticipate, but which is directly relevant to the problem we have chosen to investigate." Time and again the porcupine has forced me to look at the forest, its natural home, from a different perspective. Again and again, the porcupine has been a teacher, a storyteller of the woods, a complexifier and adorner of the world.

2 The Defense Reaction

I am driving on a country road on a May 2007 night. In the woods around me, porcupine mothers are getting ready to give birth. With their bodies demanding salt, they are converging on the richest source in the area—roadside detritus impregnated with leftover winter road salt. Cars cut them down. I stop to examine a fresh roadkill, a female with her belly swollen.

As I continue, I spot a porcupine walking the yellow stripe down the middle of the road. I pull up alongside and blow my horn to get her out of the killing zone. Instead of fleeing in the night, she whirls around, erects her quills, and comes to a full stop, daring me to come closer. Only when I yell at her does she start moving again and disappears into the roadside growth. I think she survived the weekend—I found no further roadkills on my return trip 2 days later.

The Warning of the Porcupine

In my more than 20 years of tracking porcupines in the Catskill Mountains of New York, I have come to feel that the porcupine is something of a daredevil, walking confidently through the forest, working hard to intimidate its enemies, and turning to fight only as a last resort. But when the porcupine does fight, it may so devastate its muggers that they may regret not heeding the initial warning, a warning sadly lost on speeding cars. But for its traditional enemies, the porcupine patiently repeats its warning in three sensory modalities: sight, sound, and smell.

A porcupine carries conspicuous markings of black and white visible from the back, the aspect that it tries to present to its molesters. A black line runs up the middle of the tail and expands along the lower back. The black area is fringed with white quillshafts that form a chevron on the top and extend as a

Figure 2.1. An adult male climbs a sugar maple. The black and white contrast in the lower back and head serve a warning function. (Photo by UR)

white border down the sides of the tail (Figure 2.1). The black-and-white contrast is so strong that it is visible on all but the darkest nights. It is no coincidence that skunks and porcupines use similar warning signals—they can both back up a gentle hint with a devastating follow-through.

It is also no coincidence that the warning colors both use are black and white. They are nocturnal animals, and in the dark that color combination is most easily visible. Animals marked by reds, yellows, and oranges (e.g., red efts, coral snakes, and monarch butterflies) are diurnally active and address their warnings to daytime predators, typically the birds. The porcupine's most important predators are not birds but other mammals, largely colorblind and active at night. A baby porcupine lacks the warning coloration of adults and is almost entirely black, as shown in Color Plate 2. Its quills are sharp and fully functional, but they are small and lack a powerful delivery system. Therefore, the baby must depend for protection on concealment, not deterrence. As it develops muscle power and quill size, it also develops the warning coloration of the adult. For the first few months of its life, the degree of warning-color development can serve as a rough index of the baby's age. At 3 months, the warning color is well developed.

Striped skunks and porcupines share another similarity. The white of porcupine quills, and the white stripe along the skunk's back are both fluorescent under ultraviolet (UV) illumination. The fluorescent pigment can be extracted from porcupine quills with polar solvents and can be shown to be a small molecule. (Its structure has not been determined). These fluorescent molecules function as brighteners, making the white of the quill shaft or the skunk hair a brighter white than it would be otherwise.

The porcupine's second warning comes in the auditory mode—a quiet, ominous clattering of the teeth. A porcupine launches its tooth-clack with a generalized shivering of the body, as though a chill had seized it. Then it closes its jaws, and the clattering ensues. First, the incisors vibrate against each other, then the strike zone shifts to the back, and the cheek teeth come into play. A tooth-clack episode seldom lasts longer than half a minute, but the animal can repeat the performance many times if necessary. A tooth-clack in the dark has an eerie quality. I recall an episode in our woodshed, where porcupines came to chew on a salt spill. On a May night, I picked my way softly over the grass and arrived at the open door of the woodshed, then listened in the dark. Silence. I was about to turn away when a dry, clattering sound rose up out of the back. I almost jumped. My alarmer was a porcupine, a large male, who had heard every step of my approach and had no intention of running or hiding under a log. His clattering teeth were warning me of painful consequences if I came any closer. As it happened, his threat was one I wish I had heeded, but that is a story for later. Porcupines do not tooth-clack unless an encounter is imminent. A porcupine up in a tree will remain silent even though it can see and hear a human visitor below and may even crane its neck to keep the human in view. But the same animal will tooth-clack if it wants to come down from a tree under which a human is waiting.

When the visual and auditory warnings have both failed, a porcupine falls back on its third, most immediate signal—odor. Many times, when radiotracking at night, my first indication of a porcupine only a few feet away has come from a wave of smell, pungent and unique to the porcupine. It is not the diffuse,

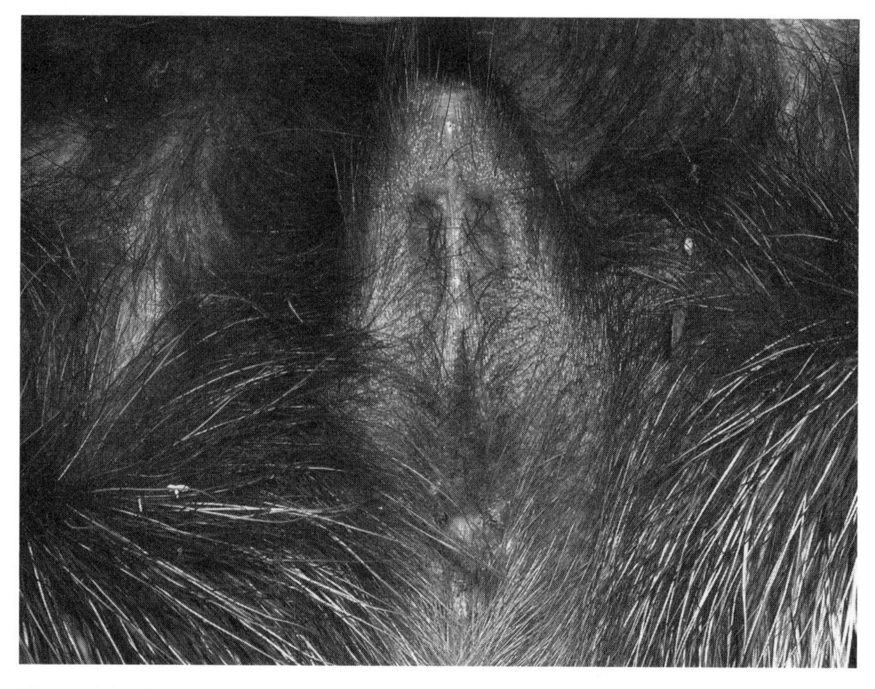

Figure 2.2. The perineal glands are paired shallow pockets lying between the penis or urethra (top) and anus (bottom). In their active state in both sexes, they produce a product disseminated by anal dragging, which appears to play a role in reproduction. (Photo by UR)

piney-ammoniacal smell associated with den sites and porcupine urine. The warning smell, released only during times of stress, is so invasive that it is impossible to stay in a closed space such as a car or small room with an aroused adult porcupine. In the open, the odor can be so penetrating that I once located a porcupine in a tree above me as I was walking on the ground below at night.

Although both sexes carry a pair of well-developed perineal scent glands (Figure 2.2), these are not involved in the production of the warning smell. The source of the odor is an area of skin above the base of the tail, called the rosette. Quills in this area are relatively short and dark-colored. Most important, the skin below is naked. In other areas of the body, the quills nestle in a layer of insulating hair. But once the quills of the rosette are erected (and the visual signal is lost), skin as hairless as that of a human is presented to view. A baby in a defensive posture displays a bare rosette in Color Plate 3. The rosette is a source of significant heat loss in the winter, as shown by the work of Clarke and Brander (1973). Up to now, the significance of the bare skin in this area has remained elusive. It now appears that the bare skin here is necessary for the rosette to function as an odor-generating organ. The evidence has been straightforward. I have taken anesthetized porcupines, lifted up the quills in various parts of the body, and smelled the skin below. Although an anesthetized porcupine is not actively generating repellent odor, the strongest smell is emitted from the rosette. Even a baby porcupine's rosette can generate repellent odor, although the smell is much weaker than that produced by an aroused adult.

The chemical nature of the odor has now been established (Li et al. 1997; Roze 2006). The odorant is a cyclic, optically active 10-carbon molecule called (*R*)-deltadecalactone. The molecule can exist in two enantiomeric states (*R* and *S* isomers), which are mirror images of one another. The *R* isomer has the smell of the angry porcupine. The *S* isomer smells like coconut. By providing an optically active warning odor, the porcupine has made its message unambiguous. Nothing else in the forest smells like angry porcupine.

Two interacting mechanisms help disseminate the warning odor. The first is the warm, naked skin of the rosette, which helps vaporize the odorant and send it free into the air. The second is the specialized quills of the rosette area. The odor is produced by sebaceous glands that surround the bases of the rosette quills. When erected, the quills squeeze the sebaceous glands and express the odorant onto the surface. There it is picked up by neighboring quills and wafted into the air. Rosette quills are specially modified to act as odor wicks—they carry a more extensive barb cover, and the barbs show more extensive overlap than on quills elsewhere in the body. The rosette quills are, therefore, classic osmetrichia, stiff hairs with erectile properties and large surface area that function in odor dissemination.

The Quill

What if the porcupine's auditory, visual, and olfactory warnings are ignored? For its own survival as much as for future credibility, the porcupine now unleashes its ultimate weapon—the quill (Figure 2.3). Upward of 30,000 quills cover every part of the body except the underparts, the muzzle, and the ears of the porcupine (Po-Chedley and Shadle 1955; Woods 1973). Quills are modified guard hairs that differ in several ways from more typical hairs. They are much thicker than normal hairs and, in both the western and eastern subspecies, are filled with a spongy matrix that makes them lightweight yet stiff (Figure 2.4). A popular misconception holds that porcupine quills are hollow. One well-known guide to mammals not only repeats this myth but also advises that, when a quill gets embedded in the flesh, the projecting shaft should be cut off to allow air to escape and shrink the quill. That advice could be dangerous. If the projecting shaft is splintered, the quill may be lost in the body. The tips of the quills carry microscopic barbs that make removal difficult once they are embedded in the flesh.

Figure 2.3. Erected quills of the back point in all directions. The quills also broadcast a warning odor. (Photo by UR)

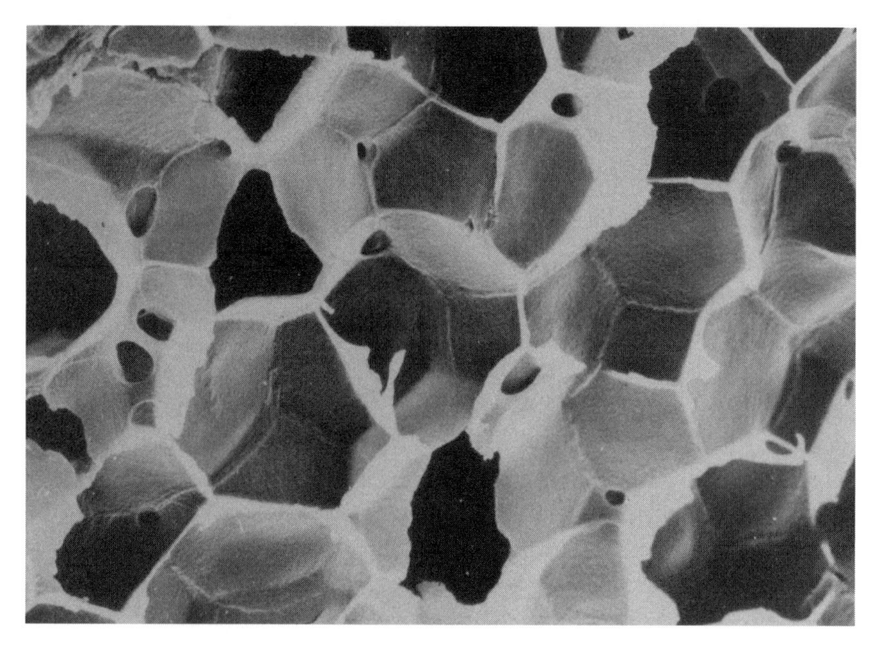

Figure 2.4. Spongy center of a quill. The spongy matrix makes the quill rigid yet lightweight. (Photo by UR)

The detachment of the quills from the porcupine is facilitated by contact with the predator (Roze 2002, 2006). A porcupine's quills are securely attached to the skin and not lost during the routine business of life such as squeezing through branches in tight vegetation. For quills to be lost and transferred to the predator, they must first be loosened. The force that loosens them is the impact of the quill tips against the predator. To appreciate why this is necessary, consider a typical encounter with a porcupine, such as one I experienced on a midsummer night in 1997. At the time, I was trying to capture the female Loretta.

In preparation, I was wearing heavy vinyl gloves to protect myself. I planned to scoop her up and place her temporarily into a snug, 3-gallon Igloo cooler, then check her radiocollar and make some measurements and observations for my research on the social structure of her species.

Loretta had other plans. She struck my glove hard with her tail. Thick vinyl stopped most of the quills, but many sharp points still pierced the fabric and dug painfully into my fingers and palm. My right hand felt useless from the pain. Loretta had won round one.

I stripped off the quill-perforated glove with my teeth and finished the capture barehanded by clapping the Igloo over the animal to immobilize her dangerous tail and lower back. Little drops of blood speckled my hand and fingers. But I had been lucky—none of the quill tips broke off to travel deeper into my body. I weighed Loretta, checked her radiocollar, noted that she was lactating, and then let her go. She moved off briskly to her baby in the woods.

But Loretta had left something of herself behind—a small forest of quills embedded in my rubberized glove. To use the glove again, I had to pull out all the quills. But when I started pulling, I was struck by how firmly the quills were anchored in the glove. So instead of just finishing the job with my fingers or long-nose pliers, I decided to measure how much force is needed to withdraw each quill.

I had an accurate spring balance, with an upper capacity of 300 g. All I had to do is attach an alligator clip to the spring and grip each quill with the clip while I gave a pull. I tallied eighty-four quills, and I measured 190 g of force per quill, on average, to extract each one of them. In fact, my result was a gross underestimate. Twenty of the quills in my glove took more pull than the balance could register. In a later experiment, I discovered the extraction tension for individual well-rooted quills can be twenty-five times higher than my first calculation, or around 5 kg apiece!

Even if the extraction force were "only" 190 g per quill, extracting all eighty-four quills at once would take a pull of more than 16 kg. That is well above Loretta's body weight—5.9 kg—and far more force than she could conceivably exert on her own, especially considering that porcupines have relatively little muscle compared with other mammals.

So how did Loretta separate herself from my glove? Not by pulling her quills out of it. Instead, she shed them from her skin. Does that solve the paradox? It might, if eighty-four quills could be removed from Loretta's skin with a force of no more than 5.9 kg—about 71 g per quill. I did the obvious experiment. During subsequent captures, I anesthetized Loretta and seven other porcupines with a quick-acting drug and measured the withdrawal tension of a few of the animals' quills. The average quill-withdrawal tension was 91 g per quill, still too much for a little animal to disengage quickly from her target. In other words, when Loretta struck my glove, she should have remained stuck to it, tied down like a bristly Gulliver by multiple tiny bonds.

The fact that Loretta was able to break her eighty-four connections in a flash suggested something wrong with my analysis. David Chapman, a histologist at Lakehead University in Thunder Bay, Ontario, who has studied porcupine skin and quill follicles through microscopy, offered a simple alternate explanation: the force needed to separate a quill follicle from the skin of the porcupine drops after the quill has been driven into an adversary. Consider one of the quills stuck in my rubberized glove. When the quill tip struck the glove, an equal and opposite reaction drove the quill shaft deeper into the skin of the porcupine. The inward push was violent enough to break some of the attachments between the base of the quill and its surrounding tissues. As a result, less force was needed to separate the quill from the porcupine's body (Figure 2.5).

How to test the hypothesis? Pulling quills from an aroused porcupine is difficult and dangerous business. Chapman suggested an elegant way around

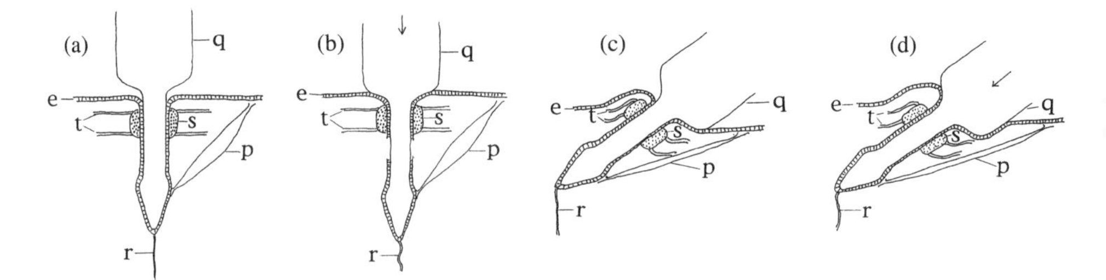

Figure 2.5. Mechanism for facilitated quill release, showing differences between erected (a, b) and relaxed (c, d) states. (a) Piloerector and transverse muscles are contracted. (b) Quill after striking an object and moving deeper into the skin, shearing its epidermal attachment and making detachment easier; deeper penetration is prevented by the spool. (c) Quill at rest, with transverse and piloerector muscles relaxed. (d) Relaxed quill after striking an object; the spool and surrounding tissues have moved with the quill root, preventing shear of root attachment. e, epidermis; p, piloerector muscle; q, quill base; r, retinaculum; s, spool; t, transverse muscle (From Roze [2000], *Journal of Mammalogy* 83:381–385)

this. Strike the back of an aroused porcupine with a block of something light and penetrable, such as cork or Styrofoam, and leave it in place. Then anesthetize the animal, separate the block from the animal by cutting off the tips of the quills embedded in the block, and measure the force needed to pull the quills with the cut-off tips out of the animal's skin.

I tried the technique on a female I named Heart. The results are clear; it takes, on average, only 54 g per quill to pull the six struck quills out of the animal. By contrast, it takes 96.5 g per quill to pull six undisturbed quills from the same area on the porcupine. Experiments with other porcupines confirmed that the tension required to extract a quill from a porcupine is reduced by about 40% if the quill is first driven into the porcupine's body. That's exactly what would happen after a tail slap or another violent contact with an antagonist.

Chapman has microphotographs that show how the trick is done. Beneath the surface of the porcupine's skin, each quill is surrounded by a spool-like structure made up of dense connective tissue. This guard spool lies just below the shoulders of the quill shaft, which flare outward sharply and so prevent the quill from being driven deep enough into the porcupine to cause the animal injury (see Figure 2.5).

When the porcupine is relaxed, the guard spools are free-floating; if you whack the quills of an anesthetized porcupine, the spools just glide in with the impact, and the quills remain anchored in the animal as firmly as ever. (That property guarantees, for instance, that a sleeping porcupine doesn't lose its quills if they accidentally press against a tree trunk.) When a porcupine is provoked, however, and its quills are erect, the guard spools are held in place by taut connective tissue in the skin. A reactive force strong enough to drive the shaft of a quill into the porcupine shears the quill's root from surrounding tissue, enabling the animal to shed its own quill and escape its injured adversary.

The porcupine uses its quills in two distinct ways: defensively, as a phalanx of spears, and offensively, as projectiles actively driven into the body of the predator. In defensive terms, the important quills are the long quills of the upper back and neck areas. At rest, they lie in neat alignment and pointing toward the back. A porcupine can groom itself by using its long claws in a front-to-back motion. During arousal, quills spring up to form a random, uplifted array of spears. Because erected quills are randomly oriented (see Figure 2.3), the porcupine is protected from all directions.[1] The longest quills on a mature porcupine may measure almost 4 inches (10 cm) long. Their shafts are white, with black tips. When the muzzle of a charging dog or raccoon passes that critical distance, the quills are transferred from porcupine to predator, sharp end out. The predator's own momentum embeds the quills.

On one June night I watched a young raccoon visiting our feeding station after a porcupine encounter. Half a dozen long quill shafts decorated its muzzle like misplaced whiskers. Because the quill tips had come to rest against the upper jawbone, the quills were fixed in place. Over the course of a week, the shafts were reduced to a ragged stubble, and the raccoon survived its ordeal. In more serious cases, a mouthful of quills may prevent feeding and cause the animal to starve to death.

The long quills produce a spectacular effect, but the short black quills on the upper surface of the tail can be far more dangerous. With a flick of the tail almost too fast for the eye to follow, the porcupine can drive the tail quills into a victim. It is probably this split-second tail slap that has given rise to the persistent but false rumor that porcupines throw their quills. Quills driven by a porcupine's tail may penetrate so deeply that they disappear under the skin and become internal projectiles.

I learned that in a painful personal encounter in September 1983 while trying to capture a 17-pound male in a beech tree, the same animal whose tooth-clacking had warned me in the woodshed. I climbed the tree and was met by powerful repellent odor and tooth-clacking. When I pressed closer, the animal's tail began to punish me, driving batteries of quills through my heavy rubber glove. I persisted. Then, with a movement so fast it blurred, the tail whipped up and hit me in an unprotected spot, the inside of my upper arm. I pulled out quills I saw sticking through the shirt, but an intense, deep-lying pain remained. On the ground, I ripped open my shirt to look at the aching spot. To my horror, I saw a large, triangular bulge under the skin. I flexed my arm. The lump, but not the pain, disappeared. The quill had burrowed into my body. I could flex my fingers but could not use the biceps. A deep, paralyzing pain stopped all attempts to lift the arm. I picked my way carefully down the mountain. On the drive home. I had to reach over with my left arm to shift the transmission. At home, the quill caused a sensation, but I firmly rejected my wife's suggestion of surgical excision. Five hours had passed since my encounter with the porcupine, and I had the definite impression that the quill had traveled downward and was now lodged somewhere inside my elbow.

Two days later, there was a strange resolution. When I got up in the morning and flexed my arm, I felt, very sharply, the presence of the quill in my forearm, ten inches down from its point of entry. In the morning light, I saw a tiny cone pushing up from beneath the skin. I flexed my arm again, and the cone disappeared. Later in the morning, I was teaching a lab. As I wrote on the blackboard, a sharp object caught my shirt arm from the inside. My shirt failed to move up and down on my arm as I was writing; then, the feeling subsided. Later, in my office, I removed my jacket and rolled up the shirt. A red spot was visible on the forearm, at the point where the quill had bulged out earlier. I felt the arm. There was no pain and no sense of a foreign body inside. The quill had emerged! I flexed and twisted the arm luxuriously. I found the quill inside my jacket arm. It was almost entirely black and looked much smaller than I had imagined it. No part of the quill was missing, but it bent sharply in the middle. It looked beautifully clean, scoured by the body to a glistening freshness. Like some mole of the flesh, it had ratcheted its way past muscles, nerves, and blood vessels to emerge far from the site of entry.

For me, the story had a happy resolution. For many animals in the wild, and for a few unlucky humans, an encounter with a porcupine may lead to permanent injury or death. Drs. H. C. McDade and W. B. Crandell, of White River Junction, Vermont, reported an unusual surgical case in a 1958 issue of the *New England Journal of Medicine*—a patient with a small intestine perforated by a porcupine quill. The patient was a 51-year-old woodsman, admitted to the hospital because of severe abdominal pain, fever, and blockage of the bowels. Surgery revealed half a porcupine quill perforating the small intestine. Escaping intestinal bacteria had set up an inflammation inside the abdominal cavity. Following massive doses of penicillin and streptomycin and a 3-week hospital stay, the man was discharged in good health. The source of the quill was described as follows. Ten days before admission to the hospital, the patient had shot a porcupine while hunting. He handled it with gloves when cutting off the forepart of the head with an ax to collect a bounty. Later, eating his lunch in a truck, he rested his sandwiches on the gloves lying on the seat beside him. It is likely that a fragment of the quill divided by the ax had stuck to the gloves and then was ingested with a sandwich.

The same article describes a less-fortunate patient. The man was admitted to the New London Hospital in New Hampshire in September 1934 with intense abdominal pain. Surgery revealed a quill perforation of the stomach and internal stab wounds from the projecting quill. The description ends, "The course was stormy, and he died on the 8th postoperative day." In 1934, no penicillin was available. Four days before admission to the hospital, the man had eaten a porcupine-meat sandwich.

In wild animals, quill injuries have been described in almost every body location: mouth, eyes, nose, lungs, intestinal tract. Some injuries lead to death or crippling, but a surprising number of the animals recover. Quills move through the body because their barbed tips pull them forward. The barbs are too fine to be seen by the naked eye, but they can be felt by the fingertip; a quill feels smooth when pushed forward through the fingers, rough when pulled back (Figure 2.6). The quills on the back of a porcupine's tail show further refinements that adapt them for traveling in the flesh. First, they are notably shorter than quills on the back and sides of the animal and consequently have a much better chance of disappearing under the skin of the victim before

Figure 2.6. Scanning electron microscope view of the tip end of a quill, showing overlapping one-way barbs. (Photo by UR)

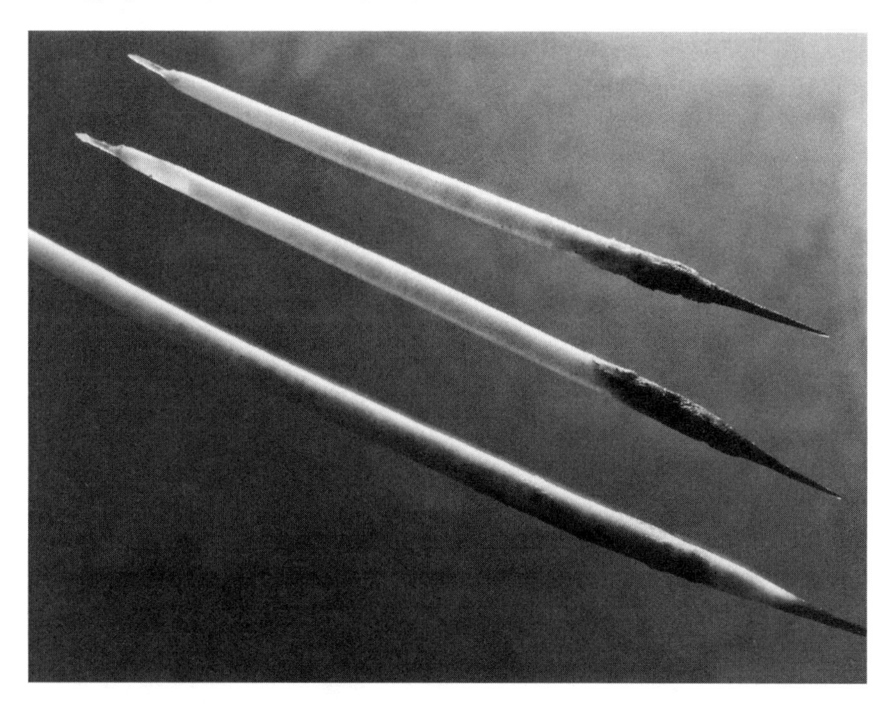

Figure 2.7. Porcupine quills with grease-coated tips. Surface grease accounts, on average, for 8% of a quill's weight and contains free fatty acids that have antibiotic properties. (Photo by UR)

they can be pulled out. Second, the tail quills are covered by a visible layer of grease, which lubricates the quills and may allow them to penetrate more deeply into the victim (Figure 2.7). And the tail houses powerful muscles for driving the quills. The base of the tail in an adult porcupine may be as thick as a man's wrist. The bones and cartilage of the tail vertebrae lie on the top, providing a firm platform for delivery.

For all those reasons, the adult porcupine has few serious enemies in the wild. To most animals of the forest, the porcupine's reputation alone is sufficient to keep it from harm. I got an illustration of that on a bitter winter night in March, when I found one of my porcupines frozen to death. The animal had been lying on crusted snow for about 2 days. A thin layer of snow had accumulated on top. I could make out, faintly, the tracks of the porcupine as it had moved forward and then lain down and fallen asleep. All around the body I found carnivore tracks, including fox, yet none had dared to touch the dead porcupine. To the little carnivores, the porcupine was inviting food. But it was, even in death, a porcupine, and therefore protected.

Quill Antibiotics

My painful encounter with the large male porcupine, so instructive in terms of quill behavior, led to one more discovery. The quill that had spent 2 days traveling through my arm left no trace of an infection. A wood splinter traveling the same path would almost certainly have produced a massive inflammation. When I mentioned it to Ward Stone, chief wildlife pathologist for New York state, he asked the logical question: Do porcupine quills have antibiotic properties?

Indeed they do. As mentioned before, the quills of the back and tail have a greasy coating. The grease layer is best developed on animals in good physical condition—that is, during summertime feeding conditions—but it does not entirely disappear even under the difficult conditions of winter. When this grease is extracted with lipid solvents and tested against Gram-positive bacteria in liquid media, the bacteria show growth inhibition. (Gram-negative bacteria are unaffected.) The extract can be fractionated. The fraction responsible for growth inhibition is the fatty acid fraction; the neutral lipids have no effect. David Locke from the Chemistry Department of Queens College has fractionated the fatty acids further into individual components and identified them by the powerful technique of coupled gas chromatography–mass spectroscopy (Roze et al. 1990). Roughly half of the total mixture consists of palmitic acid, a common saturated fatty acid of animal fat (Figure 2.8). Other major

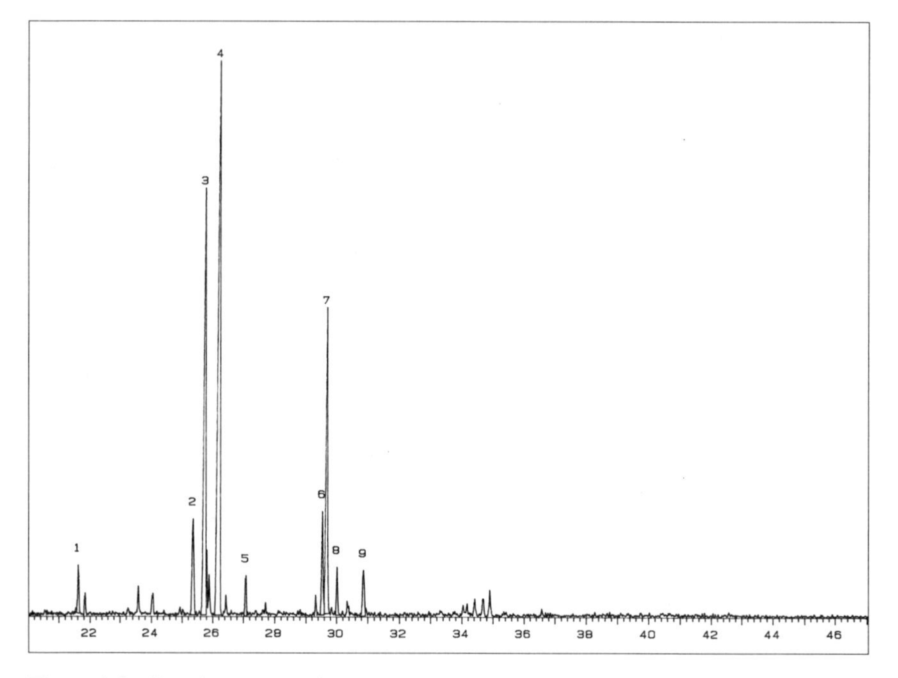

Figure 2.8. Gas chromatography spectrum of porcupine quill fatty acids. The most abundant one is palmitic acid, peak 4. (Chromatogram by David C. Locke)

components include palmitoleic, isopalmitic, and oleic acids. In addition, ten minor components account for 5% of the total. None of the fatty acids identified has proved unique to the porcupine; all are known to have bactericidal action (Wyss et al. 1945; Glassman 1948).

Human skin is likewise coated by a thin film of lipids consisting of 30% (weight basis) fatty acids (Nicolaides 1974; Lindholm et al. 1981). It has been suggested that the fatty acids on the human skin surface reduce the incidence of bacterial and fungal infections (Sauer 1980).

But the presence of substantial concentrations of fatty acids among the skin lipids is by no means common in the mammalian world. No other mammal is known to share the trait, and it is possible that humans and porcupines may be the only species to do so. Even primates closely related to humans, such as gibbons and chimpanzees, lack those components (Nicolaides 1974). (Nothing is known of species related to porcupines.) That raises the question of what the high concentrations of fatty acids are doing in these two species.

Human skin is unique among mammals in being heavily populated with sweat glands, which generate a film of moisture that might provide favorable conditions for bacterial growth. The antibacterial activity of the fatty acids should therefore serve as protection. In mammals other than humans, no film of moisture covers the skin because sweat glands are rare and find primary use as scent glands. Hence, there is no need for a defense against superficial skin infections. But the following hypothesis suggests a different role for the skin (or quill) fatty acids in porcupines.

Falling from Trees

In a deciduous forest habitat, porcupines spend a large amount of time in trees. In winter, trees are used for bark-feeding. In summer, the animals rest in trees during the day and feed in the canopies at night. Porcupines are at risk of falling out of trees because they have relatively high body weights, their foods are typically located far out on the branches, and the branches of a number of heavily used tree species are brittle. The animals that fall out of trees may suffer a variety of injuries, including impalement on their own quills. Quills that travel through the body should therefore be free of such dangerous Gram-positive organisms as *Clostridium tetani* (the cause of tetanus) and *C. perfringens* (cause of gas gangrene) (Sherris 1984b).

This hypothesis is supported by my personal observations of the arboreal life of radiotelemetered animals, for an aggregate of 66 porcupine years (Chaps. 4, 6, and 8). Additional evidence is provided by porcupine anatomy in the form of tree-climbing adaptations (Chap. 3). Also, a number of observers have given firsthand accounts of porcupines falling out of trees. Curtis and Kozicky (1944) observed a full-grown porcupine falling from a poplar tree. Marshall (1951) found a porcupine dead in a double-stemmed tamarack; the animal had fallen, wedging itself between the two stems, and had been unable to extricate itself.

I have never directly observed a porcupine falling, but the indirect evidence is compelling. In Chapter 1, I described a male that I found dead with a broken skull. His injuries could only have resulted from a fall. I have encountered five other animals who fell from trees and here describe two of them. One survived his fall, the other did not.

The survivor was the adult male Finder. At the time, Finder had worn a radiocollar for 3 years, and essentially all of his located positions had been up trees or inside dens. On June 1, 1986, I encountered the animal under a branch tangle. He made no effort to climb the adjacent trees. I captured him, anesthetized him, and made a physical examination. Finder had been gravely injured. He had suffered a blow to the jaw, and pus was oozing from the lower incisors. Two of his footpads were torn, and a laceration of the abdomen, near the left testis, was partly healed. All the injuries were consistent with falling belly-first out of a tree; none was consistent with a predator attack. But the clearest suggestion that he had fallen came from the position of the third digit of his left front limb. The animal's digit projected upward at a right angle, a probable consequence of trying to snag a branch as he was hurtling down to the ground.

Subsequently, I tracked Finder at weekly intervals and repeatedly found him on the ground. It appeared his injuries had left him unable to climb. His weight remained constant at a time when other porcupines were gaining rapidly. Finally, on July 19, I tracked Finder to the heartland of his territory and found him high up a tree. He had resumed normal tree-climbing behavior, but his third left digit remained permanently dislocated.

The porcupine who did not survive, and whom I came closest to observing in a fall, was the adult female Heart. I start tracking her on a hot and humid day in August 2003, in the company of my nephew Chris. Heart's radio signal was much weaker than the week preceding, suggesting she had either moved much higher up the mountain or was on the ground. We made our way to a high valley where stinging nettles burned our hands and a shallow, braided watercourse periodically wet our feet. Suddenly Chris cried out that he could see the porcupine—it was clearly outlined on a large, horizontal linden branch to our left. But the animal looked wrong. As I studied it through my binoculars, I confirmed it was not wearing a radiocollar. And the closer we got, the more obvious it was that the signal was not coming from above but from the ground.

As I scanned the ground, thickly covered with nettles, raspberries, Impatiens, and summer herbs, I spotted the first sign of disaster—a large broken linden branch with a mass of porcupine quills in the ground below. Perhaps 2 feet downhill, there was a second thicket of quills. These had come from Heart, whose feeding branch had broken, hurling her to the ground. She must have rolled after she fell, as suggested by the quill masses. And in a second I saw her body, downhill from the second quill mass. She appeared dead. Her back quills were pressed smoothly down; her head was under an old piece of lumbering slash. I lifted up the piece of wood under which her head was pressed and saw her head, eyes open. Her sides moved in shallow breathing.

Heart was alive but paralyzed. The porcupine in the branch above must be her baby. I reconstructed the events that had left her where she was. The night before, she had climbed with her baby to feed in the tall linden—there was a fresh niptwig directly below. The upper branches of the linden grew long and thin, with leaves at the tips. When she ventured out on one of these to feed, the branch snapped under her weight. I measured the branch at its point of breakage—it was only 1.25 inches in diameter.

Heart fell, clinging to her branch like a parachute. A young sugar maple deflected her fall, but the ground below had no thick brush to soften the impact. She crashed on rocky soil 20 feet from the base of the linden, where the base of the broken branch dug a 2-inch crater into the ground. Heart hit the ground with her back, leaving the first quill mass, rolled once, leaving a second quill mass, and came to rest 4.5 feet from the point of impact. All these events must have happened at night, with her baby on a branch nearby. It remained waiting in the tree, and when I returned to the site a week later, I discovered it had spent most of the week there, accumulating a raft of feeding niptwigs below.

There was little hope of saving Heart, but I wanted to make her final hours more comfortable—we were surrounded by biting flies. Chris and I hiked back to the cabin and returned with my granddaughter's baby bath. This served as Heart's stretcher. As I reached to touch her, she partially erected her thoracic quills, made a weak lunge at my hand with her teeth, and then subsided into a slow tooth-clacking. I removed her radiocollar, transferred her to the improvised stretcher, and started downhill. I noticed her back was strongly humped and probably broken.

Heart survived for 16 more hours at the cabin, growing steadily weaker, ignoring water and food, and losing consciousness hours before her final breath. Afterward, her skeleton confirmed the broken-back diagnosis. The vertebral column was bent sharply at the ninth thoracic vertebra, which was broken along with two adjacent ribs.

Other researchers have reported similar observations of porcupines falling out of trees. Hale and Fuller (1996) reported that 2 out of 50 of their study animals died of injuries consistent with tree falls. Both animals had been feeding in bigtooth aspen, a tree, like linden, with a limited load-bearing capacity.

Equally convincing evidence for porcupine falls from trees comes from observation of injuries. Wendell Dodge (1967), during autopsies of 200-odd porcupines in western Massachusetts, found healed fractures of legs, hips, and ribs, broken incisors, injured eyes and ears, and hernias and miscellaneous soft-tissue injuries. One animal carried an encapsulated 4-inch-long pine branch inside its abdominal cavity.

To quantify the incidence of serious injuries in the porcupine, I looked for healed fractures in 37 skeletons from museum collections. I excluded fractures of the ribs and phalanges, which are difficult to classify (Bulstrode et al. 1986). As controls, I examined 45 museum skeletons of the woodchuck (*Marmota monax*), a rodent of comparable body weight but with little tree-climbing activity. I also examined 39 museum skeletons of the raccoon (*Procyon lotor*). Raccoons climb trees readily but use them primarily for rest and denning, therefore confining their activities to the trunk and large branches. They do not venture out onto the thin, sometimes unpredictable outer branches. The results confirmed my suspicions (Roze et al. 1990). Thirteen of the porcupine skeletons (35.1%) showed healed fractures. Only 3 woodchuck and 4 raccoon skeletons showed the same. Statistically, the differences among the three species are highly significant.[2]

A list of the porcupine injuries covers every region of the body, but nine of

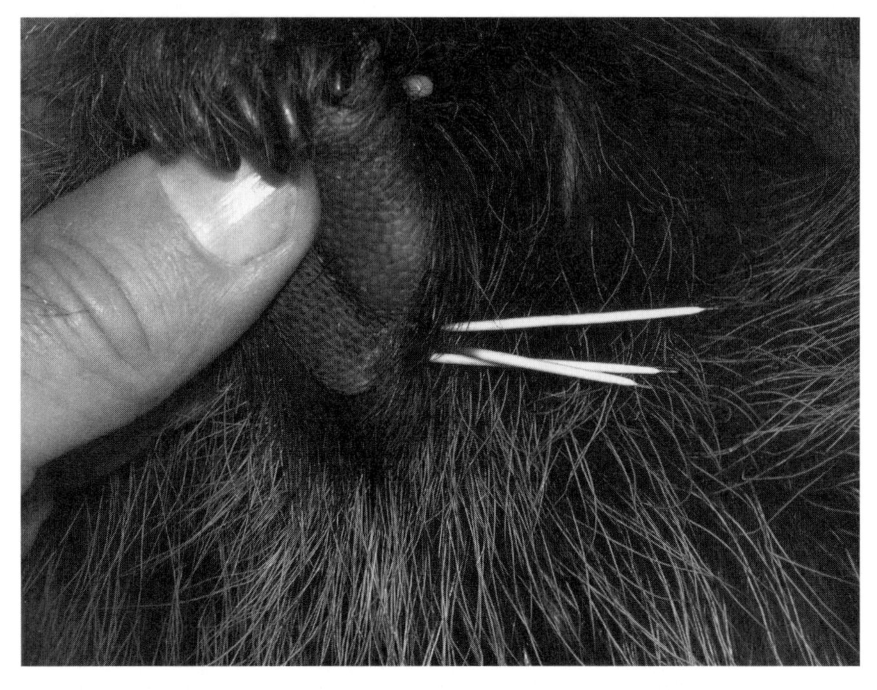

Figure 2.9. Self-quilling in a porcupine. Three quills are embedded in the heel of a footpad; an engorged tick is also attached to the foot. (Photo by UR)

the thirteen animals showed injuries of the long bones of the arms and legs; five showed injuries of the hip region (ilium, ischium, pubis, and acetabulum). Assuming that accidents serious enough to produce fractures also lead to self-impalement on quills (Figure 2.9), how effective a defense are quill fatty acids against bacteria? Much of the quill damage would be self-limited because porcupines are skilled in removing embedded foreign quills. However, quill tips may break off and travel beyond reach, as may the small quills of the tail. Under such conditions, the bacteria associated with the quill tip would be in position to start an infection. In humans, the major threats from contaminated puncture wounds are tetanus and gas gangrene, both caused by Gram-positive organisms. By eliminating that group alone, the porcupine quill fatty acids might reduce infection substantially.

Of course, the considerations that make porcupine quills safer for porcupines also make them safer for porcupine predators. Studies by Coulter (1966) in dogs and Maser and Rohweder (1983) in mountain lions show that porcupine quills cause unexpectedly few infections in those predators. But we should not conclude that porcupine quills have little effect against predators. Quills are highly effective, even against skilled predators such as the fisher (DeVos 1953; Powell 1982). But they are effective for strictly physical reasons, not because they cause secondary infections.

Predation

In its complement of quills, the porcupine possesses a formidable weapon. But how secure is a porcupine's life? The answer depends on who is in the woods with it. In early December 1978, while following a porcupine track in the snow, I discovered the animal had met a red fox (*Vulpes vulpes*). I found a jumble of tracks where the fox had circled the porcupine and the porcupine kept turning to present its tail to the fox. The evidence signaled a victory for the porcupine. The tracks of the two animals separated again, with the porcupine moving uphill, the fox downhill (Chap. 7).

It was probably a typical porcupine–small carnivore interaction. Porcupine remains have been found in red fox stomachs but only with low frequency, and it is not clear whether the fox had killed a living porcupine or fed on carrion. Other minor porcupine predators on record include the lynx (*Lynx canadensis*), the bobcat (*Lynx rufus*), the coyote (*Canis latrans*), the wolf (*Canis lupus*), the wolverine (*Gulo luscus*), and the great horned owl (*Bubo virginianus*) (Dodge 1967; Woods 1973). None may be called serious threats to the porcupine, and they may come out as losers in an attack. Ward Stone has documented a coyote that a farmer shot, fearing rabies. The animal had been moving in a circle and acting disoriented. Autopsy revealed a porcupine quill that had entered through the muzzle and traveled backwards to pierce the lower brain (Ward Stone, personal communication).

The porcupine does have a few important predators. The mountain lion (*Felis concolor*) has already been mentioned. Chris Maser's data suggest that this powerful predator feeds regularly on porcupines, making no special effort to avoid the quills but tolerating the consequences (Maser and Rohweder 1983). A striking example of mountain lion predation on porcupines is offered by Sweitzer et al. (1997). Sweitzer's study population of more than 82 porcupines in the Great Basin Desert of Nevada was reduced to less than 5 animals over a 3-year span by mountain lion predation. The authors attribute the near-extinction of their porcupines to a decline in mule deer, the lions' normal prey. But because mountain lions themselves are exceedingly rare throughout most of the range of the porcupine, an encounter between them must be considered unlikely.

The fisher (*Martes pennanti*), although a much smaller carnivore, appears to be a significant and effective predator (Figure 2.10). This relative of the weasel makes a profession of hunting porcupines, although snowshoe hares, squirrels, small rodents, and carrion are more important protein sources (Powell 1982). Fishers and porcupines are of comparable size. Adult male fishers generally weigh between 3.5 and 5.5 kg; the largest male porcupine I have encountered weighed 8.8 kg. Although porcupines do not make up a large fraction of a fisher's

Figure 2.10. Fisher in a spruce tree. The fisher's tree-climbing agility allows it to pursue a porcupine up its normal refuge. (Copyright © Wayne Lynch)

diet, they are attractive to the predator. Fishers routinely alter their winter movements to inspect active porcupine dens.

Roger Powell of North Carolina State University, who has studied the fisher both in the wild and in captivity, observed that the fisher is no more immune to porcupine quills than other animals are. The fisher kills porcupines not by tolerating their quills but by avoiding them. It attacks a vital but relatively unprotected region, the face. Powell describes a fisher's attack inside a fenced enclosure:

> An attacking fisher repeatedly circles the porcupine, attempting to bite its face. When it finds a chance, the fisher jumps in, bites the porcupine's face, and jumps back again before the porcupine has a chance to turn and strike the fisher with its tail. Repeated wounds to the face over a period of 30 minutes or more finally kill or bewilder the porcupine so that the fisher can turn it over and begin feeding on its ventral surface. Most attacks to the face are unsuccessful, and a fisher often has to check its attack after it is started because the porcupine is able to move quickly enough to protect itself. (1982, 144)

Powell believes the fisher's powerful but low-slung body allows it to attack the unprotected face of the porcupine, as opposed to its well-protected neck. In addition, the fisher's agility in trees reduces the protective value of that shelter to the porcupine.

In March 1981, I had a chance to view the results of a fisher–porcupine battle only minutes after the deed. I was visiting Wendell Dodge, the author of a classic study of the porcupine around Quabbin Reservoir, near Amherst, Massachusetts. Wendell took me out for a hike around the reservoir and showed me a cross section of porcupine dens and feeding sites. Most of the dens were empty, and the porcupine population seemed to be greatly diminished. One active den was inside a hollow white oak, about 3 feet in diameter. Patches of snow still lay on the ground, and Wendell followed a faint feeding trail to a hemlock grove 50 feet away. Under the hemlocks, he knelt down, stared at something on the ground, and called out in his laconic New England way, "Here's something that will interest you." I hurried over. Lying on the ground was a freshly killed porcupine, its body still warm. Our arrival had interrupted the killer, who could be identified from the method of attack—the porcupine's head was missing. This was the attack mode of the fisher. The head had been peeled out of its skin, with tufts of hair, quills, and small patches of skin piled next to the torso. After consuming the soft tissues of the head, the fisher would work into the body. In other fisher kills, the inner tissues are consumed through a hole made in the chest or abdomen. The dead porcupine had been a yearling female. She was not pregnant, and she appeared to be in excellent physical condition, sleek and well fed.

According to Wendell, the fisher is rather common in the area, so much so that local trappers poached on it. (Because the Quabbin Reservoir is part of the Boston city water supply, hunting and trapping are not permitted.) It is clear that the fisher is an effective predator on porcupines. Two approaches have been used to gauge its importance in controlling porcupine populations. First, porcupine population densities in a given area have been compared before and after the introduction of fisher. That requires serial data spanning several years. The second approach is a parallel comparison; porcupine densities are measured in similar habitats, one with fishers, the other without.

The serial approach was used by Powell and Brander in 1975, with additional data extending the study to 1979 (Powell and Brander 1977; Powell 1982). Their study site was the Ottawa National Forest in upper Michigan. Before the 1960s, the fisher had been extirpated from the area. Between 1961 and 1963, Michigan introduced 61 fishers from Minnesota into the Ottawa National Forest. The release has been successful. It appears that the porcupine population in the national forest has declined severely following fisher introduction (Table 2.1).

The parallel study was carried out by Earle and Kramm (1982) in winter 1976. They compared a 919-ha (3.5-square-mile) area of the Ottawa National

Table 2.1. Porcupine population density in Ottawa National Forest

Year	Porcupines per 100 km^2
1962	1180
1970	510
1971	400
1972	440
1973	610
1974	520
1975	400
1976	280
1979	80

Source: Data from Powell (1982)

Forest with a 644-ha (2.5-square mile) area in the Hiawatha National Forest. Both sites are in upper Michigan and have a similar tree-species composition and topography. They differed in fisher abundance. The fisher was well established in the Ottawa forest thanks to the 1961–1963 releases; fisher was rare or absent in the Hiawatha forest 120 miles to the east. Porcupine population densities for the two areas were 40 animals per 100 km^2 in Ottawa and 350 per 100 km^2 in Hiawatha. Earle and Kramm's figures for Ottawa are much lower than Powell and Brander's 1976 data for the same area. Powell and Brander took a census of a smaller area (178–283 ha); the difference probably reflects local variation. It is clear that in the fisher-stocked forest, porcupine abundance is significantly reduced.

Another serial study based on telemetry and capture-recapture methods shows a porcupine population driven to near-extinction by predators, primarily fisher (Mabille, Descamps, et al. 2008). The study area encompasses 200 ha of forest and open fields in the Parc National du Bic in Quebec, Canada. Fishers arrived in the area in 1999. Between 2000 and 2006, essentially all the porcupines in the area were captured and fitted with radiocollars or other forms of identification. Radiocollars were fitted on 97 porcupines for a total of 64.6 porcupine-years of telemetry. A starting population of 117 animals in 2000 had declined to only 1 animal by 2006. The authors recovered 60 carcasses, excluding 10 whose deaths were either study-related or statistically out of bounds. Of the remaining 50, 28 died by predation, 13 by winter starvation, 3 by falling out of trees, 2 from vehicles (roadkills), and 4 of undetermined causes. Among predation victims, 86% were killed by fishers and 14% by coyotes. Deep snow increased predation risk.

My own experience with fisher-porcupine interactions has been broadly similar. Fishers were reintroduced into the southern Catskills by the New

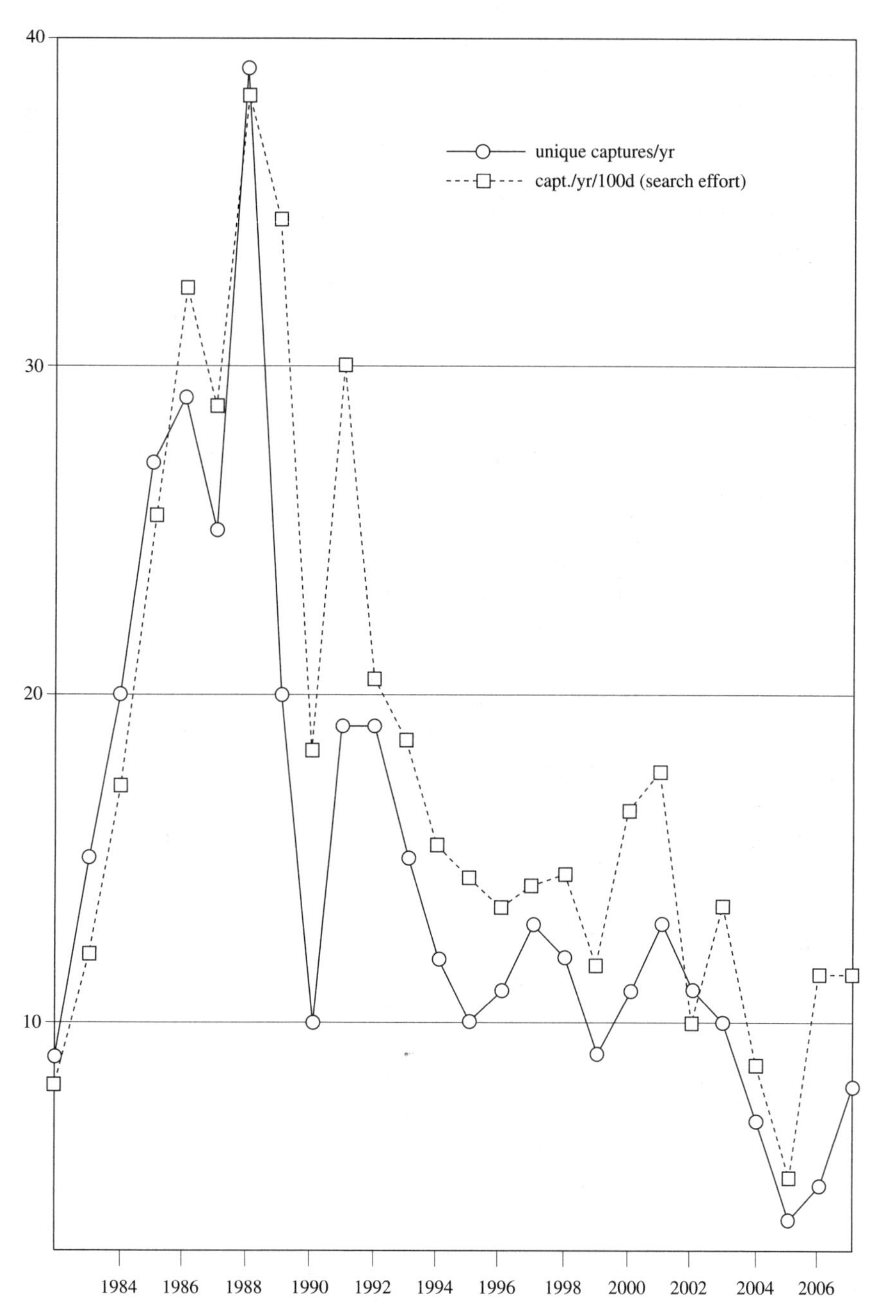

Figure 2.11. Porcupine study population 1982–2007. Between 1987 and 2005, the population declined by 86%, most likely because of fisher predation. The initial increase from 1982 to 1987 is ascribed to the ending of a scabies epidemic.

York Department of Fish and Wildlife in 1976–1979 (Wallace and Henry 1985). They reached my study site in the northern Catskills a decade later, as shown by snow tracking. (The snow tracks of fishers are unique in terms of pattern, size, and the presence of five toes.) There followed an immediate and severe reduction of the porcupine population (which I measured as the number of unique animals captured per calendar year, equalized for search effort). By 2007, the porcupine population stood at roughly 23% of its 1988 peak (Figure 2.11). The reduction of the porcupine population has been accompanied by near-total disappearance of the snowshoe hare, another prey favored by fishers.

Unfortunately, the link between fisher presence and porcupine decline in my study is far from air-tight. I have not been able to find the kind of physical evidence of fisher predation that I found at the Quabbin Reservoir in Massachusetts or that Mabille, Deschamps, and colleagues (2008) found in Quebec. The reason is straightforward—I scaled back my radiotelemetry program after 1989, with only one to three radiocollars deployed at any one time. Therefore, most of the porcupines that died or disappeared after this time did not carry radiocollars, and their remains were never found. Yet the presence of fishers after 1989 appears to be the only factor that distinguishes this period from the one preceding it. What accounts for the 1982–1988 rise in population? As I explain in Chapter 11, I believe this represents the recovery of the population from a scabies epidemic (see Figure 2.11).

For the 18 carcasses I did find between 1982 and 2006 (most prior to 1989), winter stress (animals found frozen in their dens after losing over 30% of their body weight) accounted for six, human predation (four hunters, one homeowner) for five, injury (all fallen from trees) for four, disease (both scabies) for two, study-related (one an anesthesia accident and the other caught on a branch by its radiocollar) for two, and unknown causes for three.

When I started my study more than 20 years ago, I thought of the porcupine as leading a carefree life, backing up its confident advertisement with an impregnable defense. But the compelling message of all long-term studies is that the porcupine defense is effective only against unspecialized predators. When faced with a fisher, a mountain lion, or a human being, the porcupine becomes a classic prey animal, more vulnerable than most because it is so poorly equipped for flight.

Notes

1. The quill erection mechanism has been studied by Chapman and Roze (1997). Quills in both the thoracic and rosette regions are arranged in clusters of three to five. Histology reveals

different quill-erecting mechanisms in the two regions. Rosette quills are erected by two layers of piloerector muscles that insert in the dermis of the skin. Both thoracic and rosette quills, in addition, carry a complex system of transverse and oblique muscles associated with the roots of each cluster of quills.

2. $\chi^2 = 11.66$, df=2, $P < 0.005$.

3 Anatomy

Acidie was a porcupine I knew for only 6 months, during the summer and fall of 1985. She died on December 27, on the sixth day of winter, but her free life in the woods had ended months earlier.

When I first capture her in June 1985, she is a young mother working to raise her offspring; she is chewing plywood scraps in our woodshed to replace the sodium that she is losing in her milk. After radiotagging her and releasing her in the woods, I realize she is behaving differently from my other porcupines. Almost every time I locate her, she is on the ground, not up in a tree. Most often she is hiding under the green tangles of raspberry and blackberry or under the slash left by lumbermen. Sometimes she slithers between thick ground vegetation to find shelter in an available rock den, unusual behavior for a porcupine in the summer. Because I find her so often on the ground, it is easy to capture her and get week-to-week physical data. Again a strange situation reveals itself—Acidie is gaining no weight as the summer progresses, a time when porcupines build up body fat for the winter ahead.

On September 22, I pick up her signal once more. I have spent the day walking through green woods flecked with the first yellows of autumn. Acidie is my last animal of the day; I catch up with her in a mature woods, where the forest canopy closes over bare ground beneath the trees. With no blackberry tangle to hide in, she is attempting a normal escape—climbing a tree. I catch sight of her 8 feet off the ground, clinging to a young linden trunk. I stop 30 feet away to study her. Acidie is pressed uncomfortably against the trunk, moving neither up nor down, her limbs spread wide. When I move closer, a gesture that would send an ordinary porcupine scooting up the tree, Acidie attempts to climb but is defeated by a small branch that blocks her path. Standing directly below, I hear a light sound, like raindrops falling. It is her tail bristles slipping as she starts to slide backward. I grab her.

After anesthetizing her, I learn that her weight is no longer holding constant but has begun to drop precipitously, slipping from 4.0 kg in June to 3.4 kg in September. She should be gaining weight on the year's rich supply of acorns and residual linden leaves, but she is unable to climb trees to reach those important porcupine foods. She has been surviving on ground vegetation, and the quality of that resource is declining. Acidie shows other signs of debilitation: the grease layer on her quills has disappeared, and she is heavily infested with lice. She appears to have poor bladder control. She will die if left alone; I intervene.

In New York City, I have Acidie examined at the animal care facility of the Bronx Zoo. My initial diagnosis of a urinary tract infection is ruled out by normal hematology and urine values. There seems to be no reason to take an X-ray, the one procedure that might have shed light on her condition. And she is examined under anesthesia, eliminating a second vital clue. Thus, the examination only deepens the mystery.

I install Acidie in a cage in our basement and feed her a mixture of apples, rodent chow, and water. She shows excellent appetite but puts on little weight. Over the weeks, the mystery of her condition is gradually unveiled. She lacks all sensation in her tail and hindquarters; I can touch them and get no response. Touching anywhere else produces quill erection, tooth-clacking, and angry screaming. A lower spinal cord injury would explain the numbness and loss of bladder control. She retains some motor control. Her hind legs move in a robot fashion, but gradually that ability diminishes and she drags herself about the cage with her front limbs. Then she somehow gashes the side of her tail. The wound spreads and festers and does not respond to antibiotics. She dies of the infection. An autopsy reveals the underlying cause of her misfortunes. Her pelvis has been crushed in a fall, injuring her spinal cord. The ischial bones have knitted together in an abnormal position; the muscles of her tail and hind legs have atrophied; the bladder is empty.

The wreckage of a once-living body presents a sad picture. Only her front limbs recall some of the music of her existence. The muscles of the forearms, layered thick as a carpenter's, convey a raw power. They alone had pulled her partway up the tree before I captured her and had propelled her through the underbrush for most of a summer. Even in death, the front limbs retain the beauty of function well executed.

General Anatomy

The anatomy of the North American porcupine, *Erethizon dorsatum,* is a composite of general traits shared with all members of its taxonomic group, the hystricomorph rodents, and more specialized (derived) traits that may be shared only with other New World porcupines or may be restricted to *Erethizon* alone. The porcupine's broadest affinities are with the order Rodentia, and they are broad indeed—some 42% of the roughly 5400 mammal species recognized today are rodents (Wilson and Reeder 2005). Rodents possess ever-growing incisors, allowing them to exploit foods that require gnawing. (The German word for rodent, *Nagetier*, means "gnawing beast." The English word has a similar root, being derived from the Latin *rodere,* "to gnaw.") They have no canine teeth, which restricts all larger rodents to a herbivorous mode of life. Between the incisors and the cheek teeth lies a toothless space, the diastema, into which the lips can be drawn to facilitate gnawing. The lips can be closed behind the incisors to seal the mouth to unwanted fragments dislodged by the incisors. The cheek teeth often have complex grinding surfaces, an aid in the grinding of plant materials. The grinding motion can be front to back, side to side, or both.

Most rodents also have a free (unfused) radius and ulna (bones of the lower arm). Plantigrade locomotion (walking on the soles of the feet) gives most of them a stable but not particularly speedy travel mode (Macdonald 1984). They are mostly small animals. The largest, the capybara (*Hydrochaeris hydrochaeris*) of Central and South America, may weigh up to 66 kg (146 pounds). But a 4 million-year-old fossil rodent from Uruguay, *Josephoartigasia monesa,* fed on aquatic plants and weighed 1700–3000 pounds. The porcupine is close to the high end of today's rodent body-weight spectrum. In North America, only the beaver surpasses it. In my study, 54 males over 1 year old averaged 5.53 kg (range 2.4–8.8 kg), and 82 females of similar age averaged 4.59 kg (range 2.2–7.2 kg). The difference between the sexes is significant.[1] Porcupines from the Idaho shrub desert appear to reach higher body weights. Craig and Keller (1986) report an average weight of 11.5 kg (25.3 pounds) for five adult males.

Within the order Rodentia, the porcupine belongs to the suborder Hystricomorpha ("porcupinelike forms"), which includes not only the New World and Old World porcupines but also such New World families as the guinea pigs, chinchillas, agoutis, and spiny rats and the Old World gundis and mole rats (Wilson and Reeder 2005). The hystricomorphs are characterized by a unique positioning of the chewing muscles (the masseter muscles). One arm of the masseter passes through a large opening below the orbit called the infraorbital foramen. That configuration increases the efficiency of chewing. The hystricomorphs are also reproductively specialized (see Chapter 12). Females tend to have long gestation periods and produce a small number of young. In males, the penis carries a sacculus urethralis and is normally retracted within the penis sheath. In the words of Stuart Landry, "It is depressing to note how many male hystricomorphs court by urinating on the female and indeed, often on their helpless offspring" (1977). In females of most species, a vaginal closure membrane is in place at all times except at estrus and parturition. During pregnancy, the ovaries develop accessory corpora lutea that secrete the large amounts of progesterone needed to support a long pregnancy.

That is the body plan the North American porcupine has modified to fit its own way of life. *Erethizon* feeds largely in trees. Its body is therefore modified for climbing and maneuvering in trees and for feeding on leaves, fruit, and bark. It shares those adaptations with its Central and South American relatives of the genera *Coendou, Sphiggurus,* and *Echinoprocta.* A set of adaptations that it does not share with its southern relatives is the ability to survive under conditions of extreme cold. Most porcupines of the world, both Old World and New World forms, are animals of the tropics. *Erethizon* is the only porcupine able to survive up to the tree line in the far north.

Arboreal Adaptations

To get to its food, *Erethizon* must climb. The claws of both front and hind limbs are extremely long, fitting like curved conical hats over the outer segments of each digit. There are four claws in the front, five in the back (Figure 3.1). A vestigial thumb is visible on the skeleton, and in some individuals a tiny thumb projects to the outside. The claws serve several functions. Most important, they fit into crevices in tree bark, working just like the leg hooks used by telephone linemen. On smooth-barked trees, the progress of a climbing porcupine can be read from residual scratch marks. The claws, which are long enough to reach through the quill layer to the skin below, are also used as a comb in self-grooming. In addition, the front claws can fold in over the palm to manipulate foods such as apples or small branches, and they are used to snag tree branches to bring them into feeding range (see Figure 6.5, p. 95).

The porcupine's limbs show another tree-climbing adaptation: the palms of the hands and the soles of the feet are naked and have a pebbly texture that suggests crepe rubber. These tuberosities increase friction against the tree trunk. When trunks or branches are too thin to climb with its claws, the porcupine can use its palms and soles alone; its powerful clenching thigh muscles allow it to clasp a small tree trunk by sole friction, leaving the front limbs free for food-handling or exploration (Woods 1972; McEvoy 1982). Footpads also help the animals sense and conform to the branches that they traverse during nocturnal feeding expeditions. Footpad size increases with the weight of the animal. Because adult males tend to be larger than females, they generally have larger footpads.

Figure 3.1. Four digits on a porcupine front paw. The naked footpad provides traction against trunks and branches; the curved claws sink into crevices in tree bark. (Photo by UR)

One more important piece of tree-climbing equipment is the tail. The tail of the porcupine is a truly versatile organ—I have already discussed the use of the upper, quill-studded surface in personal defense. The lower surface is unquilled but bears instead stiff, backward-pointing bristles embedded in a deep layer of connective tissue. When this surface is pressed against a tree trunk, a technique shown in Color Plate 4, the bristles engage the bark and prevent downward sliding. In descent, the tail becomes an exploratory appendage, lifting up and down at each claw-hold and testing the way below.

An observation in the summer of 1983 illustrated how important the tail can be as a tree-climbing aid to the porcupine. In the middle of a July night, I captured a small male that had been sitting immobile for at least an hour in a sweet-apple tree, a great favorite of porcupines. He made no movement as I approached and put up no struggle when I grabbed him. I named him Moribund Porcupine, or MP for short, and held him in a cage overnight to examine him by daylight. I discovered then that his left rear leg was paralyzed, perhaps damaged in some old fall from a tree. He climbed the sides of the cage by using only three legs and walked by throwing out the bad leg like a prosthesis. The leg had been paralyzed for some time; the left footpad was narrower than the right, and its claws were long and sharp, not worn down by climbing. I kept the animal several days for observation, then attached a radiocollar, and freed him on a fine summer afternoon. He set off at a brisk but swaying pace, walking on three legs and holding the bad leg off the ground. I followed at a distance. He crossed three stone walls and investigated several trees along the way before choosing one to climb—a tall, medium-size linden.

How does a three-legged porcupine climb a tree? By using its tail as a fourth leg! Instead of directing his tail vertically, he swung it to the left to oppose the traction of the active right leg and gain an effective four-point purchase. The most difficult maneuvers involved climbing over projecting branches by laborious lifting of the bad leg over the obstacle. But again the tail helped out, becoming a leg substitute. MP did not stop climbing until he reached open sunlight atop the canopy. As I watched through my binoculars, he began to feed slowly, eating leaves directly off the branch he was clinging to, until I could see a row of bristling petioles. He then climbed even higher, for tastier leaves. The branch bent ominously. A small wind picked up, and the top of the once-vertical branch bent to the horizontal before rebounding. As the branch and its clinging porcupine swung in great arcs, MP went on feeding. Happily, I renamed him. No longer would he be MP, the Moribund Porcupine; he was now Imp, Improved Porcupine.

A porcupine's movement up or down a tree is a close-contact experience. The animal spreads its limbs wide and drags its belly along the trunk. There are two interesting consequences to this mode of tree climbing. First, the animals carry no quills along the belly and inner surface of the limbs. Any quills growing there would be scraped off by the intimate contact with tree bark. Second, male porcupines do not carry an external penis. It would suffer continual abrasion as the groin was dragged up and down tree trunks. The penis is normally retracted inside a cloaca-like structure, a solution made possible by retractor muscles attached to a tiny bone inside the penis, the baculum. The arrangement is part of the animal's hystricomorph heritage. The female's external reproductive organs are also hidden through most of the year. The vaginal opening is normally covered by a closure membrane. Whatever the original function of the adaptation, it serves to keep bark chips and debris out of the body while the female is crawling up or down tree trunks. There is thus a logical consistency to the body architecture of the porcupine, with all anatomical arrangements contributing to efficient tree climbing.

One final tree-climbing adaptation may not be as satisfactorily resolved; it relates to body weight. As John Eisenberg (1983) has pointed out, there are upper limits to the size of arboreal leaf-eaters, dictated by the mechanical problems of small branches, where the leaves grow, supporting animal bodies. Porcupines may be pushing at that limit, at least when feeding in the more brittle branches of deciduous trees.

Digestive System

Porcupines climb trees primarily to feed on leaves or bark. In this, *Erethizon* has few mammalian competitors, because few other North American mammals (and no widespread species) possess the two traits necessary for that way of life: the ability to climb trees and the ability to digest leaves or bark. Only a few fellow rodents, wood rats of the genus *Neotoma* and tree voles of the Pacific Northwest, can match those traits, but they are smaller and their distribution is more limited.

Feeding on leaves is a difficult occupation because tree leaves, during most of the growing season, contain only low levels of nutrients such as protein, sugar, and starch. The bulk of the leaf biomass consists of complex molecules not accessible to mammalian digestive enzymes. This is dietary fiber, containing diverse chemical fractions such as celluloses, hemicelluloses, lignin, pectin, cutin, and suberin. Celluloses and hemicelluloses, although resistant to mammalian digestive enzymes, can be digested by enzymes secreted by certain bacteria. The porcupine incorporates such bacteria as essential parts of its digestive system (Balows and Jennison 1949). Because the bacterial enzymes work slowly, the porcupine digestive tract is very long and has a slow passage time. It is a major organ system in the porcupine body, making up 26% of total body weight (Table 3.1). Compared with humans, the porcupine devotes

Table 3.1. Weights of major porcupine organ systems

Organ	Percentage of porcupine body weight
Skeleton	13
Digestive system	26
Muscles	26
Skin, quills, fur	27
Heart and lungs	3
Miscellaneous	5

far more of its body resources to digestion and far less to skeletal muscles or to the heart and lungs.

Much of the bacterial activity of the digestive system takes place in a large sac, the caecum, at the junction of the small and large intestines. The porcupine caecum is about the same size as its stomach. It is here that a broth of digestive bacteria interacts with finely ground plant materials under anaerobic conditions and for extended periods of time (Vispo and Hume 1995). In the end, much of the plant fiber is fermented into small molecules—acetate, propionate, and butyrate—that are absorbed from the caecum and large intestine and used to supply the animal with energy. Johnson and McBee (1967) estimate that on average 16% of the animal's energy requirements are supplied by the porcupine's caecal fermentation. In some individuals, the figure is as high as 33%. The authors point out that this contribution would be most important during the periods of poorest diet, when snow covers the ground and the only food available is low-quality tree bark.

Unlike rabbits and beavers, who also have large caeca, the porcupine is not normally coprophagous (it does not re-ingest feces for a second round of nutrient extraction).[2] That means that the bacteria in the caecum are not themselves used as food sources. The porcupine's relatively long large intestine, which serves to absorb the fermentation products, partially compensates.

The large intestine is efficient in resorbing water from the intestinal contents (Vispo and Hume 1995). I have found that porcupines in the summer can obtain all their water requirements from the leaves they eat. In my study area, most high-mountain water sources such as seeps, pools, and rivulets dry up from mid-July to September, yet my radiotelemetered animals do not migrate to the valleys at that time. I have further documented the absence of drinking by maintaining porcupines on tree leaves alone for 5-day periods. They had no interest in the water I offered afterward.

Animals that practice extreme water conservation, such as hibernating black bears, invoke a second mechanism of water conservation besides intestinal resorption—urea recycling (Thornton et al. 1970; Bauchop 1978). Intestinal bacteria break urea down to ammonia, which is then used to synthesize protein. This takes a load off the kidney function and hence reduces water loss through urine. Whether porcupines are able to recycle their urea deserves experimental investigation.

Before the enzymes and bacteria of the digestive system can do their work, the food must be reduced to a fine, almost dust-particle size. Porcupines carry three kinds of teeth well suited to the tasks of scraping, gnawing, and grinding rough foods—four incisors, four premolars, and twelve molars (Figure 3.2). The teeth are symmetrically placed in each quadrant of the mouth, hence the dental formula reads (I 1/1, C 0/0, P 1/1, M 3/3)$\times$2=20 (where I=incisor, C=canine,

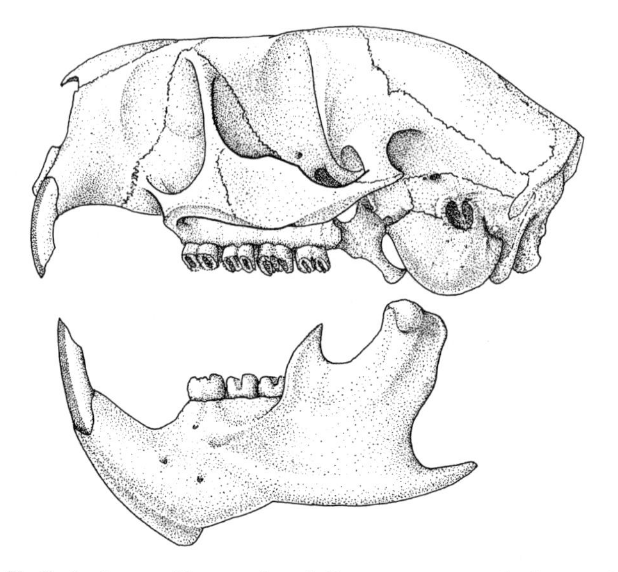

Figure 3.2. Skull of a 1-year-old porcupine. A diastema separates incisors and cheek teeth. While all cheek teeth have erupted, the deciduous premolars will be replaced by larger permanent premolars at 2 years of age. (Drawing by John Winsch)

P=premolar, and M=molar). The formula represents the teeth of the upper and lower jaws on one side of the mouth, times 2 for the total for both sides.

Porcupine incisors are the longest teeth in the set. They are faced with a bright orange enamel that contains iron salts, which function as hardening agents. Because the outer surfaces are harder than the inner ones, they can be sharpened to produce a chisel-like occlusal edge on each tooth. The sharpening is done by the reciprocal action of the opposite incisors (Macdonald 1984). The chisel edges are excellent tools for scraping bark and for nipping small twigs (required for leaf-feeding). No amount of scraping or nipping can wear out the incisors because, as in all rodents, they grow continually from deep inside the jaws. Although each incisor loses 100% of its length to wear in a year's chewing, its length always remains the same. The thin layer of enamel reveals a further level of structure when a cross section is studied under a high-powered microscope—the enamel is built up from crystalline prisms. In porcupines as in most other rodents, the enamel contains several layers, with prisms in adjacent layers oriented in different directions. This plywood principle of enamel architecture gives it great resistance against cracking and chipping (Koenigswald 1985).

A porcupine's mouth has two chambers. The incisors sit in the outer chamber, or vestibule. When one examines a porcupine in a tree through binoculars, the animal appears to be grinning, its large orange incisors exposed. In fact, the lips are closed behind the incisors, across the diastema (the toothless stretch of jaws between incisors and premolars). The arrangement allows the animals to rasp off the outer bark of trees in the winter while keeping their inner mouths closed and warm.

Bark or leaves are ground to a fine consistency by the cheek teeth—the premolars and molars inside the inner chamber—which work like millstones. In hystricomorphs (although not in all rodents), the jaw motion is front to back, not sideways as in our own mastication. The cheek teeth have a level, slightly rippled grinding surface, resulting from vertical folds of hard enamel, that is, the grinding surfaces have alternating hard and soft domains. Porcupine cheek teeth, unlike the incisors, have fixed roots. Once worn down by a lifetime of chewing, they cannot be replaced. In very old animals, the cheek teeth are ground down close to the pulp and have lost most of their occlusal relief. When the process continues, there may be exposure of the pulp cavity and decay may set in. Porcupines are long-lived animals; the limit on their life expectancy may well be set by erosion of their grinding teeth.

As in all mammals, a porcupine's tooth complement changes with age. For the first two years of life, tooth complement alone gives a measure of age (Dodge 1967; Sutton 1972). A porcupine is born with eight teeth: four incisors and four deciduous premolars are fully erupted. During its first summer (by 4 months of age), its jaw elongates to accommodate additional teeth. It adds a set of first and then second molars. Early next summer, when it has finished its first year of life, it adds the third molars (Figure 3.2). The adult tooth set is reached a year later, at age 2. At that time, the deciduous premolars are lost and replaced by permanent premolars. Because the permanent premolars are larger than either the deciduous premolars or the molars, they are easily identified in the mouth.

Beyond 2 years, accurate ages can be obtained by the technique of tooth sectioning (Earle and Kramm 1980). A cheek tooth is extracted from a dead animal and soaked in dilute acid to remove calcium. What remains of the tooth is its collagen matrix, retaining both the original shape and inner structure of the tooth. The matrix can be sectioned and stained to reveal cementum growth rings, one for each year the tooth has been in place. Thus a 4-year-old animal will show four growth rings in its first and second molars, three in the third molars (which grew in a year later), and two in the premolars (which grew in at age 2). The incisors, growing continuously, show no rings.

Winter Adaptations

During Catskills winters, while bears and woodchucks sleep below the ground, the porcupine remains active in the snow. Its species range extends to the tundra line, and in places porcupines have been reported living on the tundra itself,

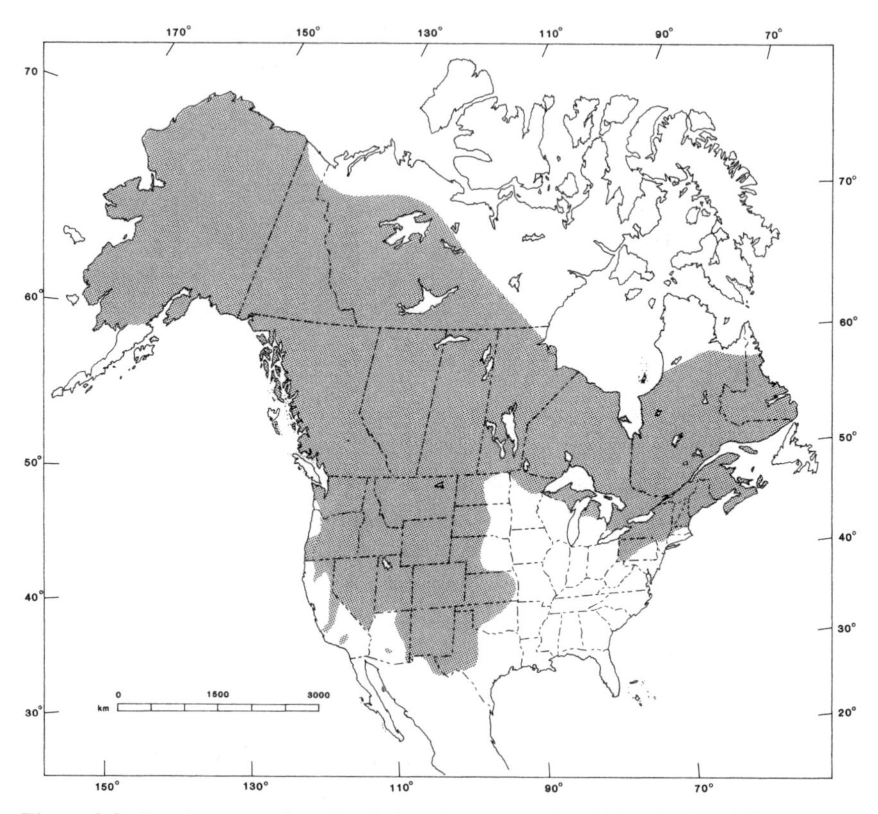

Figure 3.3. Species range of the North American porcupine. (After Dodge 1982)

a remarkable performance for a South American immigrant (Figure 3.3). Part of the porcupine's winter adaptation is behavioral. It spends much time in a den (abandoning it after snowmelt); it reduces travel and moves through the snow in its own tracks or under rock overhangs. But for winter survival, porcupines are equally indebted to anatomy. They prepare themselves for the coming winter from midsummer onward. Most important of their preparations is replacement of the underfur they have molted at the beginning of summer.

Porcupines grow five types of hair: the quills, bristles on the undersurface of the tail, vibrissae (whiskers), guard hairs, and fur. The guard hairs, which resemble long, thin, barbless quills, help shed rain and sleet and convey tactile information about the immediate environment (Po-Chedley and Shadle 1955). The fur, however, constitutes the animal's primary heat shield, helping to maintain a temperature differential between +37 °C inside the body to as low as –35 °C outside. The same differential exists between a glass of boiling water and a room-temperature table below it. In the inorganic system, the water soon cools down, the table warms up, and the differential is diminished

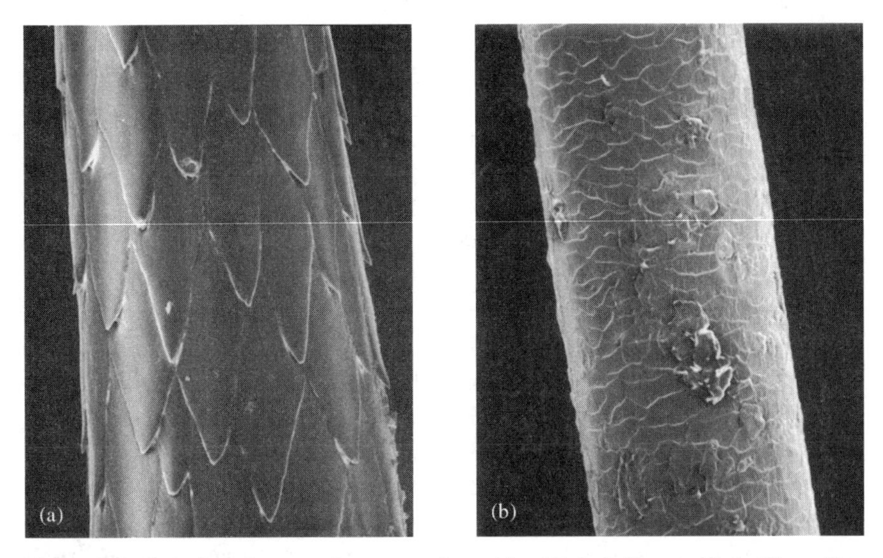

Figure 3.4. Scanning electron microscope views: (a) quill shaft (b) guard hair. The quill barbs are responsible for the quill's one-way movement through tissue. X200 (Photo by UR)

or lost. But in the porcupine-snowbank system, the temperature differential remains undiminished for weeks, perhaps months, of hard winter. Such a performance must begin with good insulation, in this case the porcupine's pelage of fur, guard hairs, and quills (Figure 3.4). It is interesting that the porcupine can control these components separately. When temperatures drop, porcupines can press their larger quills against the body, while piloerecting the fur and guard hairs (DeMatteo and Harlow 1997). This mixed pelage of the porcupine has a high insulation efficiency—67% better than predicted for an animal of this size (Bradley and Deavers 1980). The fur is renewed each fall, growing back after the spring molt. It is so thick that it hides all but the long rosette quills. Porcupines in the winter have a fluffy, deceptively cuddly appearance. The illusion is enhanced by their tendency to roll up into a ball when resting on a high tree branch. I see them often in the branches, when I walk the mountains on a calm winter day. They are always alone in the tree, lying still and without sound, and symbolizing the dark, banked-down spirits of winter.

As mentioned in Chapter 2, one area of the porcupine's body is sparsely furred even in winter—the rosette (see Color Plate 3). Steven Clarke and Robert Brander (1973) have made infrared measurements of surface body temperatures of porcupines on winter nights. On the night of February 28, 1969, near Amherst, Mass., when the outside air temperature measured –14 °C and the "sky temperature" was –32 °C, an adult female in an open field showed the following surface temperatures: back –13 °C, side –11 °C, rosette –7 °C. That is to say, the animal's back was so well insulated that it measured only marginally warmer than air temperature (although still considerably warmer than the heat sink of the sky). The rosette was 6 °C warmer than the back, indicating increased heat loss. The porcupine reduces rosette heat loss by pulling this area of skin down toward the tail and keeping cold at bay with the more thickly furred areas of shoulders and mid-back. The reason for the slight warming of the sides is that the fur is shorter there than on the back, especially as it tapers down to the belly.

One reading that Clarke and Brander did not record was the thermal emission from the naked and relatively large footpads. How do porcupines avoid heat loss from these potential radiators? The footpads of porcupines in the winter feel cold to the touch, much colder than the belly. Other northern animals with warm bodies but cold feet, such as the snow goose, have a countercurrent heat exchange mechanism. Small arteries, carrying heated blood to the feet, run adjacent to small veins, carrying cold blood back to the body. As the two vessels continue their parallel paths but countercurrent circulations, heat from the arteries radiates out to warm the returning blood in the veins. The process is so efficient that when arterial blood reaches the surface capillaries in the feet, almost no heat is dissipated because it has all been returned to the body. It seems possible that porcupines have evolved a similar mechanism.

But porcupines are not perfect thermos flasks—they do lose heat to the environment. At external temperature estimated at –2 °C, resting porcupines cannot maintain body temperature through normal basal metabolism (Fournier and Thomas 1999). They increase their metabolic rates, the equivalent of turning on the furnace on a cold winter night. Using the tail when a leg is paralyzed, scraping bark when tree leaves have all fallen, turning up the metabolic rate when insulation fails—these are porcupine responses to adversity. Perhaps it is that flexibility of response, not a perfected body plan, that has allowed the porcupine to expand its species range in a continent it invaded only 3 million years ago and to hold it almost everywhere the forests grow.

Notes

1. $t=3.40$, $\mathrm{df}=134$, $P<0.01$.
2. Porcupines may practice a limited form of coprophagy to acquire appropriate intestinal microflora (see Chap. 13).

Plate 1. Porcupine quillwork: detail from Sioux arrow case. (Photo courtesy River Trading Post)

Plate 2. Mother and baby. Although the baby has tiny quills below its guard hairs, its all-black coat lacks a warning signal and is best suited to concealment. (Copyright © Wayne Lynch)

Plate 3. Musa in defensive posture. Her short baby quills leave most of her rosette skin bare. (Photo by UR)

Plate 4. Adult male climbing, tail pressed against the bark. (Photo by UR)

Plate 5. A porcupine at the water's edge. The animal's buoyant quills and good swimming skills allow it to harvest sodium-rich aquatic vegetation. (Copyright © Wayne Lynch)

Plate 6. Musa in linden tree, blocked by a small branch above her. (Photo by UR)

Plate 7. Musa in plum tree, confident of her climbing skills. (Photo by UR)

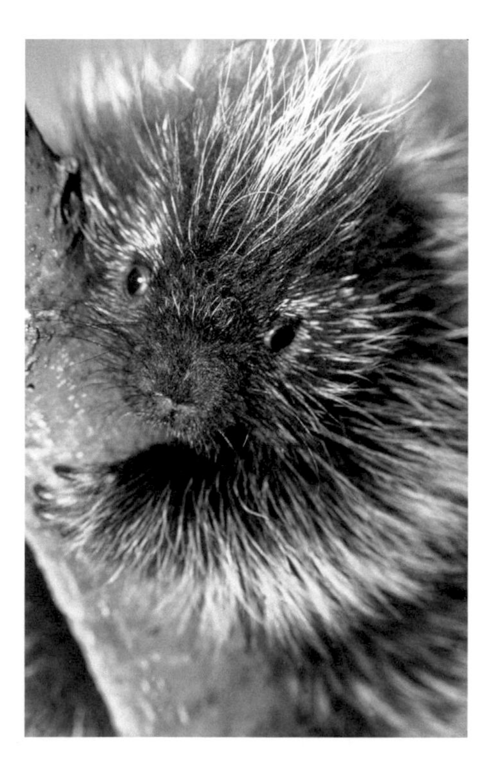

Plate 8. Bus, Squirrel's first baby, shortly before dispersal. (Photo by UR)

4 Spring Foraging

The winter is over. In the second week of April, I spot a drop of lemon yellow on the trail, a round-leaf violet in first bloom. I am sweeping through the valley, tracking Rebecca but not believing my antenna. Rebecca's radio pulse is coming from the wrong side of the Roarback; I suspect a reflected signal. As I spiral in on her, passing an open Canada anemone and a stand of Carolina spring beauty, I find myself in an unfamiliar part of the valley, a corner of our own woods that I have seldom explored.

Rebecca announces her position from a distance; her dark mass hangs so far out in the small outer branches of a tree that it seems to float in air. The tree is a tall, vigorous sugar maple. As I nail a small plastic identification label to the tree, I notice that Rebecca has begun to move. She is not trying to climb higher—there is nowhere higher left to go. She has begun to feed.

I lie down under the tree to watch. Periodically the dark body elongates and two arms appear, then a branch is gathered in and the body resumes its rounded form. Through the binoculars, I see Rebecca has nipped off a good-size terminal branch and is holding it in her paws. She is nibbling off the maple buds, turning the twig over and over in her paws. This is a new mode of feeding. The previous week she was still in her winter mode, concentrating on hemlock and on the bark of sugar maple. This week the maple buds have burgeoned large enough to be worth harvesting. As Rebecca turns the twig in her paws, she sometimes abbreviates it by nibbling off a spur. A few twigs have already been processed and thrown away; all remain snagged high up in the canopy. But now she nips off a small piece, which starts its way down, bouncing from branch to branch in the sugar maple yet always managing to break free and struggle lower. Freed of the maple, it hits the top of a second-story beech below it. The twig is almost snagged and hangs, undecided, for a second before its last free fall. I examine it. It has been carefully processed. The smallest twigs have been nipped off in their entirety; the large buds along the sides have been excised. Only a few small buds on the inner surfaces of branch forks remain. I silently admire Rebecca's workmanship.

I look again at the trees and sky above me. The vertical masses of tree trunks are holding up the vast lacework of bud-swollen branches. The evenly spaced buds of beech and sugar maple, dark knots against the sky, generate a large-scale rhythmicity. The red maples are in flower, a red smoke rising in the canopy. It is the end of day. The air is absolutely still. High above, I hear the light rasping of the porcupine gathering her spring harvest. For her, as for many animals that feed on the forest, the period of greatest abundance is just beginning.

Plant Chemistry

The Catskills forest is largely deciduous. With few exceptions, the trees shed leaves in the fall and enter 6 months of winter dormancy. Porcupines, who feed on the bark and twigs of the dormant trees, slowly starve. Their weight drops steadily, their fat reserves become depleted, and they reach a low point of existence. Most of the natural deaths in my porcupine population occur during winter. This long famine is followed by the richest feast of the year as trees pump nutrients out of deep storage and into swelling buds and unfurling leaves. Major changes are generated in plant tissue chemistry.

Figure 4.1. Rebecca foraging in outermost branches of a sugar maple. (Photo by UR)

The concept of a changing plant tissue composition flies in the face of our deeply ingrained expectation of constancy in the body tissues of animals. Regardless of what animals may eat, how much they exercise, or what the time of year may be, there is little change in the balance of major ions and organic molecules in their blood and internal tissues. The same is not true of plants. The cell sap of trees may undergo violent fluctuations in nitrogen, sodium and potassium, hydrogen ions, and a variety of complex organic molecules. Those levels may fluctuate from season to season, from day to night, and from one part of a plant to another. It is the genius of the porcupine to become aware of the changes and to exploit them for its own needs.

I do not know how porcupines learn that the April buds of a sugar maple have suddenly become more nutritious than the maple bark they were gnawing a week before, but it is clear that animals failing to show such discrimination would not survive. It is remarkable not only that porcupines are able to track the changing resources available in the forest but also that they appear to do so almost instantaneously. Springtime changes in the sugar maple illustrate some of these principles. While snow still covers the ground and buds are still tiny, the turgor pressure in the sapwood begins to rise on the warm, sunny days that follow cold nights. Catskill farmers know it as the running of maple sap. The sap is collected in buckets, then boiled down to syrup in the sugar house. Porcupines do not appear to share the farmers' excitement. They continue feeding on the inner bark, and the tree may bleed along the edges of chewed areas. But the maple tree is changing under the influence of its internal sap flow. At first the roots grope deeper into the ground, searching out nutrients and water. Then buds begin to swell as leaf primordia expand inside. At the same time, the nutritional quality of the buds is rising.

Nitrogen Supply

For porcupines, as for most herbivorous animals, the crucial dietary component is nitrogen (Mattson 1980). Nitrogen does not supply metabolic energy in the same sense that carbon does. Instead, it is needed to build proteins (muscle, collagen, body enzymes), nucleic acids, and a range of important small molecules. The bulk of our dietary nitrogen comes from meat products, from eating the bodies of other animals. Because porcupines feed on plants exclusively, they depend on them for nitrogen. But plants are generally poor nitrogen sources.

Nitrogen content is typically expressed in terms of crude protein level. The crude protein content of porcupine winter foods (bark and evergreen twigs and needles) ranges from 2 to 6% in the Catskills (compared with 100% for lean meat). That is not enough to replace the normal wear and tear of body tissues, and porcupines lose more nitrogen in urine and feces than they gather in their diet. Their body weights decline.

The swollen buds of sugar maple show nitrogen levels that are, comparatively speaking, stratospheric. The maple buds collected by Rebecca average around 22% crude protein (dry weight basis), more than that of most enriched breads and cereals. The twig tips carrying the buds average only 8.5% crude protein. By excising the buds and throwing away the twigs, Rebecca was showing a fine discrimination in food quality.

Catskills porcupines consume other foods with properties similar to sugar maple buds. A food much preferred by Rebecca has been the buds of linden trees. The linden is a rare tree in the mountain forests; hence, its contribution to the porcupine diet is lower than that of sugar maple. But because the linden is intensively browsed in both spring and summer, the trees often take on a shorn look. I have tracked Rebecca to lindens where the ground below was littered with twigs whose buds had all been surgically excised. Although linden buds show lower crude protein content than sugar maple buds, they are available earlier in the year.

Another spring delicacy is the catkins of quaking aspen. Catkins emerge before the leaves do, and by late April the aspens are hung with fluffy, pendulous wind-pollinated flowers that are a promise of brighter green to come. Quaking aspen is a sun-loving tree and is not found in mature forest. The porcupines must therefore travel to find the trees on the edge of forests and in abandoned clearings. They feed as they do in maple and linden trees, by nipping off terminal twigs, harvesting the catkins, then discarding the twigs. Because aspen branches are brittle, the porcupines nip them closer to the trunk. Aspen niptwigs can be quite large. Again, nitrogen analysis shows a big difference between the tree parts harvested and the parts rejected. Catkins contain 20% crude protein, the rejected twigs only 7%.

During early spring, some porcupines continue feeding on hemlock twigs and needles, a winter food. The porcupines that do so have left their dens early and are using hemlock habitat for shelter; the edible portions of hemlock average only 8% crude protein. Hemlock drops out of the diet entirely by early May.

Phenolics

Porcupine feeding choices change rapidly during the months of April and May. One week all animals may be feeding on the buds of sugar maple. Fourteen days later, not one animal can be found on that diet; the forest has turned green with leaves. With bud-break, new opportunities open at the same time that old ones are foreclosed. In the case of sugar maple, the buds represent food, leaves poison. Once sugar maple leaves are fully open, porcupines climb the trees only to rest, never to feed. The difference between the buds and leaves has little to do with crude protein content. Rather, the difference reflects a plant defense against its predators.

In the 1970s, Paul Feeny (1970, 1976) of Cornell University showed that young oak leaves defend themselves against winter moth caterpillars by secreting high levels of defensive compounds called tannins, complex molecules composed of phenolic subunits. Their biological effectiveness results from their ability to complex with proteins (Harborne 1988).

Because sugar maple and oak share many aspects of life history, I wondered if the maple's defense against porcupines might not follow the same strategy. I collected thick spring buds, extracted the tannins in hot methanol-water, and measured total phenol levels (a measure of tannin content). The test tube contents turned a pale blue, indicating tannins were present but in low concentration. Two weeks later, I collected young leaves from the same tree and repeated the measurement. The extract turned as dark as ink. By the first week of May, the phenol levels had risen from less than 2% (wet weight basis) in buds and emergent leaves to 6.5%, a level intolerable to porcupines. This is far more rapid than the rise of tannins in oak leaves.

The maple-porcupine-tannin story has a further twist. The Catskills forest contains four species of maple. The two most abundant species (and the two achieving the most massive sizes) are sugar maple and red maple. My students have trouble distinguishing between them. There are subtle differences in leaf shape and bark texture, but in the summer it takes an experienced eye to tell a red from a sugar. Yet for porcupines, the difference is absolute. In early spring, sugar maple buds form their most important food source. The red maple, either in bud or in the leaf stage, is never touched. The difference, once again, appears to be the tannin content. Sugar maple is undefended in early spring, but red maple is at all times chemically well armored. Its buds, twigs, and leaves are always heavily laced with phenolics, and porcupines respect the difference.

Beech

With the disappearance of sugar maple from the list of spring foods, my porcupines rapidly move to alternate food sources. I discover this one mid-May while radiotracking along a hillside. The sun is setting. Along the ridgetops, the forest still stands bare, but here at lower elevations, a fresh green adorns most trees. Ahead of me, Squirrel's signal grows stronger, and then I see her through the undergrowth. She is up a wrist-diameter beech tree and barely 12

Figure 4.2. Squirrel feeding on emergent beech leaves. (Photo by UR)

feet off the ground (Figure 4.2). The tree is kinked at the top, and she rests on an S-shaped twist in the trunk. I stand below and look up. The little beech tree carries young leaves. Canopy-height beech above it are still in the bud stage. In the background, a towering red maple reddens the sky with emerging leaves. Squirrel stares down at me silently. Although she now remains motionless, the evidence of her activities lies under my feet. The ground is littered with fresh niptwigs, their proximal ends cut at an angle and the leaves chewed off except where a twig has fallen out of the animal's grasp. I taste a few of the leaves. They have a faintly astringent, salad flavor. When I bring the leaves into the lab to measure total phenols, I find a value below 1%, well within the porcupine tolerance range.

I reflect on the porcupine's powers of discrimination. It has had to realize not only that beech trees are coming into leaf this week but also that the beech trees in leaf are the understory saplings, not the adults in the canopy above. And, because most porcupine feeding takes place at night, the distinctions have to be made in the dark.

Ash

Two or three weeks later, another spring food enters the porcupine menu—the young leaves of white ash. Ash leaves are among the last to emerge in the spring; once they have done so, the forest canopy closes and spring wildflowers melt away. Ash quickly displaces beech as the major porcupine spring food, though beech consumption continues at a lower rate for several weeks. Part of the reason may have to do with the ease of climbing the two trees. The smooth bark of beech places it among the most difficult trees for porcupines to climb; porcupines almost never climb beech trees to rest. Ash trees, on the contrary, have a finely furrowed bark that offers excellent claw-holds. Ash trees also have much larger leaves than beech. A single compound ash leaf may weigh 3–4 g, a single beech leaf perhaps one-tenth that amount. Ash leaves are low in tannin content, less than 1% total phenol.

Ash leaves, like all other leaves, are harvested by the niptwigging technique. Ash niptwigs are striking; they look like green-quilled stick-figure porcupines. Each twig is surrounded by a bush of long green petioles (leaf-stalks). To determine why foraging porcupines reject the petioles, I compared the tannin and nitrogen content of the leaf blade, petiole, and twig. Tannin levels did not differ significantly from part to part. But leaf parts collected in June showed marked differences in crude protein levels: leaf blades 16.1%, petioles 5.5%, terminal twigs 7.0%. It is clear that porcupines feed only on the leaf portions richest in protein, a selectivity made possible by the careful manipulation of individual niptwigs. The rejection of petioles has proved to be a general feature of porcupine tree-feeding; in every case I have examined, the petioles contained less nitrogen than the leaf blades. Howler monkeys (*Alouatta* spp.), which occupy the leaf-feeding niche in Central and South America, show a similar antipetiole bias in at least some of their feeding.

Ground-Feeding

A number of authors report porcupines feeding on ground vegetation during spring and summer months. Taylor (1935) reports that in the coniferous forests of Arizona, ground feeding made up 40% of the diet in the spring and 85% in summer, as determined from stomach contents. Identified items included dandelion, goldenrod, sedges, clover, and raspberry. Graham W. Smith (1982), working in a northeast Oregon Douglas fir–juniper forest, found porcupines feeding in the spring on succulent ground vegetation. When the groundcover dried out in July, porcupines returned to their winter diet of tree bark and evergreen needles. One interesting case of ground-feeding comes from the desert country of Colorado and Arizona, as described by Julio Betancourt, Thomas Van Devender, and Martin Rose (1986). Here the porcupines feed on grasses and such flowering herbs as thistle, goosefoot, bindweed, and clover. The plant remains have been found in their droppings.

Ground-feeding represents a geographical variation in feeding behavior. Neither deserts nor evergreen forests present the springtime feeding opportunities available in the mixed hardwood-hemlock forests of the Catskills. The only ground vegetation I have observed my porcupines feeding on consistently has been raspberry leaves. Like all other preferred spring foods, the raspberry is a low-tannin, high-protein food. It makes up a relatively small fraction of the spring food of free-ranging porcupines; however, captive porcupines will eat armfuls of offered raspberry canes. Ion, a giant male and the first porcupine I radiocollared, spent several weeks in our basement until a radiocollar could be shipped. Ion adapted well to a diet of rodent chow and water, but his great love was apples (which he took from my hand) and early-May raspberry canes. Twice a day, I brought in a load from our garden sufficient to crowd his spacious cage. Twice a day, I returned to pull out defoliated stems. In the confines of the cage, the porcupine's feeding technique did not show the neatness observed in the wild, but it was clear that he was again avoiding the petioles. Ion ate up so many of our raspberries that our summer crop was significantly reduced.

At least three physiological consequences follow from the rich spring diet of porcupines. First, they put on weight at the most rapid rate seen during the yearly cycle. Second, they molt their winter fur, retaining only guard hairs, quills, and tail bristles. Third, the spring diet leads to sodium loss and, hence, a sometimes fatal lust for salt. The last phenomenon is explored in Chapter 5.

Weight Gain

Human beings, accustomed to a relatively constant body weight year-round, may be surprised by the large annual weight swings of wild animals. An adult British hedgehog starts life in spring weighing about 600 g (1.3 pounds). After a summer's feeding, it has ballooned to 900 g, a 50% increase (Morris 1983). Essentially all of that weight will be lost again during winter hibernation. Free-ranging bull elk from Alberta have spring body weights of about 200 kg (440 pounds) and reach 300 kg by late summer (Gates and Hudson 1979). Then they begin to lose weight because of the stress of the breeding season. The loss continues during winter starvation.

Swings in porcupine body weights are less extreme, although still large compared with humans. Adult females gain approximately 25% during the nonwinter feeding period (April 15 to October 15). The weight gain, however, is not evenly distributed over the summer, as illustrated by the 1983 weight data for the porcupine Squirrel (Table 4.1). From April 30 to June 10, the ani-

Table 4.1. Weight changes in female porcupine, 1983

Date	Weight (kg)
April 30	5.0[a]
June 10	6.3
July 10	5.6
August 21	6.2

[a] 11 lbs

mal was gaining weight at the spectacular rate of 0.63% per day, corresponding to an annual increase of 230%. Much of the increase was pumped into a growing fetus; Squirrel gave birth on June 12. The weight drop observed on July 10 was caused by the loss on parturition. Despite the drain of lactation, Squirrel continued gaining weight in July and August, almost reaching her weight of June 10. The midsummer weight gain was accomplished at the slower rate of 0.11% per day, corresponding to an annual increase of 40%. The difference is accounted for by the drain of lactation and the lower quality of summer forage.

Molting

Like the spring weight gain, molting is another mammalian phenomenon without a close counterpart in human experience. And like spring weight gain, molting in the porcupine is diet-controlled. Not all mammals go through a synchronized molt. Hedgehogs, the porcupines' spiny look-alikes, molt hair and quills gradually, a few at a time (Morris 1983). Likewise, humans replace their hair gradually, a few strands at a time. Not so the porcupine—it may molt all its winter fur (but not its quills or guard hairs) in one burst in early spring.

I encountered the male Finder on one such occasion, in early May. The trees were still bare, but maple buds were thick and ready to burst. As Finder's signal pulled me toward the ridgetop, I encountered glades of wildflowers. There were green expanses of wild garlic, toothwort, dark-leaved blue cohosh, and yellow round-leaf violets, along with the larger Canada violets, white on top and purple below. But Finder was not wandering among such splendors. He had been feeding on sugar maple buds (Figure 4.3). Under the tree I found a prodigious amount of niptwigs, most with buds excised but a few still intact, having slipped out of the animal's grasp. I looked up. Pressed vertically against a branch, his body was surrounded by an irregular halo of loosed fur. Every once in a while, a tuft broke loose and floated away on the breeze. A

Figure 4.3. Finder sampling sugar maple buds in early spring. (Photo by UR)

week later, the molt was finished. Finder looked sleek and glistening, clothed now only in quills, guard hairs, and tail bristles.

Other porcupines were molting at the same time. Ten hours after I had watched Finder's fur blowing like a dandelion clock, I captured Eight, a medium-size male chewing for salt in a woodshed. Great projections of hair gave his body a ragged outline, and in crevices of the woodshed I found tufts of shed fur, lost while the porcupine was squeezing through. The molt continued through my next encounter 6 days later. A spring storm had driven Eight into a ground den. In the entrance, I found masses of molted fur, plus a dozen quills, presumably loosened by his long claws as he combed himself. Far from the male's ground den, I located my old female Squirrel. She had been sheltering through the storm in a hemlock tree. Under the tree lay a mat of porcupine fur, evidence of the ongoing molt.

All three animals were molting in the first and second weeks of May. The snow pack had melted a month before, and rising temperatures would make any winter fur superfluous. But two animals, captured in March of the same year and kept at home, began molting in the last week of that month. Had they done so in the wild, they would have risked freezing to death or exhausting their already depleted body reserves needed to drive metabolic heat production. The first animal captured was Cameo, an adult female. On March 10, in deep snow and intense cold, I found her in a moribund state. She allowed me to remove her collar without anesthesia, indicating an alarming decline of the will to live. I decided to gather her in till she rebuilt her strength. A day later, I encountered a baby porcupine in a foot-deep open rock crevice. It carried no radiocollar, but an eartag identified it as Topmost, a male born the previous May. Its weight was barely 4 pounds and its strength so low that I decided to gather it in along with Cameo. The two animals adapted well to captivity and to their diet of apples and rodent chow. Both put on weight. On March 23, 12 days after his capture, I found Topmost sitting on top of his food bowl like a teddy bear, paws folded in front. His appearance was somewhat spoiled by stray tufts of fur emerging from his body. On following mornings, the floor of his cage was matted with squashed fur tufts. On March 27, the molt peaked. I had to clean his cage twice a day and collected enough fur to stuff a small pillow. His body had an irregular outline as fuzzy locks raised the contour. He was losing almost no quills or guard hairs. By March 29, his molt was finished. No more hairballs appeared in his cage, and he was looking sleek and well groomed. But now Cameo, sitting in an adjacent cage, began the process. Her hair tufts first appeared on March 28, peaked on April 1, and were substantially finished by April 4. Both animals finished molting within 6 days, roughly the same time span shown by animals in the wild.

While the animals were captive, I was running Kjeldahl nitrogen analyses of their droppings to get a rough indication of their nutritional status. At the time of capture, both sets of droppings were averaging 6% crude protein, typical winter levels in starving animals. Within 7 days, crude protein levels in their droppings had jumped to 20%, a value normally reached only during mid-May feeding. The nitrogen status of molting animals in the wild shows the same picture. Droppings collected from molting free-ranging porcupines show crude protein values between 15% and 23%, suggesting that in the wild as well as in captivity the spring molt of porcupines is induced by a rise in dietary nitrogen. The picture is complicated by an underlying biological clock. Porcupines gathered from the wild and fed high-nitrogen diets in December do not molt.

If a rise in dietary nitrogen induces the spring molt in porcupines, it constitutes an ecologically well-designed trigger. In many mammals, the spring molting trigger appears to be an increase in day length (Ebling and Hale 1970). That might not be a reliable signal for porcupines, who show considerable environmental variability in den-quitting times. A ground-denning animal that is active nocturnally may see little daylight. On the other hand, the high-nitrogen pulse of spring foods represents in itself a climatic, as opposed to a calendar, signal. The emergence of the buds and leaves of spring reflects local growing conditions. For example, sugar maples on top of Vly Mountain (altitude 3530 feet) experience bud-break 2 weeks later than the same species in the valley below (altitude 1500 feet). The locations differ not in day length but in average temperature and snowfall, the very factors with which the winter fur must cope.

Late Spring Diet

The spring diet of the porcupine continues to evolve. Sugar maple buds, crucial in early spring, drop out of the diet with bud-break. The leaves of beech and ash, which filled the gap left by the loss of sugar maple, decline more slowly and are replaced by true summer species. The reasons for the decline of beech and ash remain to be elucidated. For beech, the total phenol levels gradually climb to 2% (wet weight). Perhaps more important, the succulent young leaves turn tough and papery. Whereas a young beech leaf tastes little different from tart lettuce, the mature leaf has the quality of parchment. The decline of ash has no such obvious basis. Its leaf tannin content, whether measured as total phenol or as condensed tannin, is among the lowest of all forest trees. It is possible that ash develops other defenses that eliminate it entirely from the porcupine's diet by mid-June.

All trees on the porcupine's spring and summer diet, however, have one potent and unbreachable defense against herbivores. They disrupt the leaf-eater's sodium metabolism, driving porcupines to chew on barns, cabins, and human implements—a topic explored in the following chapter.

5 Salt Drive

On a mild May midnight, I am driving north through the Notch, a 3.5-mile stretch of highway with hills rising so steeply on either side that no houses or hunters' cabins punctuate the wilderness. Suddenly I am braking for a small gray shape out of the night. A porcupine, with unhurried dignity, is crossing the road. Down the highway, a less lucky animal lies with quills scattered, hit by a passing car. Driving slowly now, I spot two more porcupines apparently feeding by the road. Then another roadkill; then two more active animals. I am astonished. In 3.5 miles, I have encountered five active porcupines. During most of the year, a single porcupine by the road would be noteworthy.

I turn and head back—I want to see if my passage has scattered the animals, and I am curious about their activities. As I reach the first one, I pull off to the side, leave the car's headlights on, and get out for a look. The animal is chewing something, but as I get close, it drops the object and moves back into the shadows. I pick up a small piece of deadwood clearly showing the porcupine's tooth marks. Deadwood contains no food calories for porcupines. The animal has been using the wood for a different purpose—it has been salt-feeding. The Notch is heavily salted by road crews during the winter. The last snows have disappeared a month ago, and spring rains have washed most of the salt out of the soil. But porous deadwood along the roadside has held some salt and attracted the porcupine. I confirm the presence of salt later by laboratory analysis.

Retracing my path along the Notch, I count not five but six active animals. One is licking a wet spot in the middle of the road; the rest are active along the roadside. At the head of the Notch, I pull over to write down my numbers and to recover from the shock of seeing so many porcupines in one night. A car pulls up alongside. A red light begins to flash—the state patrol. An officer leans out and asks, "You have any trouble?"

"No," I answer, and add, "There are a lot of porcupines on this road. Two of them got hit."

The officer looks at me sharply. "You all right?"

I see I have made the wrong impression but blunder on, "I'm just trying to say to drive carefully."

"Oh, sure," the officer says, and speeds off. I follow him at a distance and count only four porcupines on my third pass along the road.

The Notch remains an excellent place to watch porcupines on spring nights because the surface and shoulders of the road represent the single most concentrated salt source in the region. Farther north, in the old farming country of Lexington and Westkill, more attractive salt sources abound: farm outbuildings, cabins, and human implements. There the animals are safe from speeding cars but face new dangers from aroused property owners who may resent having their cabin walls or truck undercarriages chewed.

Salt-seeking is not a behavior restricted to porcupines. The urge is observed in all large and medium-size herbivores living in salt-depleted environments. It has been studied in deer, moose, elephants, woodchucks, fox squirrels, mountain goats, domestic sheep, and cattle, to name only a few. Nor is salt drive restricted to modern herbivores. It was present in sometimes-fatal form in the mastodons and mammoths of the Pleistocene. A spectacular collection of Ice Age elephant remains at Big Bone Lick, near the Kentucky-Ohio border, holds the remains of animals trapped in a salt-impregnated quaking bog (Teale 1965). They died 10,000 years ago, attracted by the bog salt and foundering on the unstable bog surface.

The metabolic basis of this ancient and deeply ingrained herbivore habit is tied to the beginnings of life. The first cells are thought to have evolved in a saline solution, and the interior of every cell today remains a salty environment. For animal cells, that is also true of the exterior of every body cell, but the salt components inside are different from the components outside. Salts inside the cells are largely potassium-based, the salts outside largely sodium-based. The opposed ionic gradients across the cell membrane set up an electric potential that is crucial to the function of nerve and muscle cells. During the action potential of nerve or muscle, sodium ions rush into the cell and potassium ions escape, causing a reversal of the electric potential. Without an excess of sodium ions outside the cell and potassium ions inside, a nerve cell cannot send messages and a muscle cell cannot contract. In herbivores, muscle and nervous tissues must maintain a 1:1 ratio of potassium to sodium ions. (Whole-body ratios tend to be closer to 2:1 because intracellular volume is larger than extracellular volume.) But herbivore diets, consisting only of plant tissues, diverge widely from that ratio. Plants contain no nerve or muscle and need only maintain the saline environment inside the cells. The result is that plant tissues contain potassium-to-sodium ratios as high as 500:1 (Weeks and Kirkpatrick 1976, 1978). Herbivores feeding on plant tissues are therefore flooded with unwanted potassium, which must be excreted to maintain internal balance. Because sodium and potassium are closely related chemically, the kidneys may not be able to excrete the excess potassium without loss of essential sodium as well.

If an animal's potassium-to-sodium ratio rises significantly above 1, severe consequences may be expected. Perhaps the most dramatic example is the lethal injection used in some criminal executions. The injection is a cocktail consisting of a muscle relaxant, a nervous system depressant, and potassium chloride. The potassium salt stops the heart immediately. For animals in the wild, milder symptoms of a sodium-potassium imbalance appear well before that point. The animal shows apathy and lassitude as nerve and muscle functions decline, followed by weight loss as the body tries to raise its plasma sodium concentration by excreting water. Blood pressure falls, circulatory abnormalities develop (including edema of the brain with associated central nervous disturbances), and death follows.

To forestall such disasters, herbivore bodies tighten up hormonal mechanisms of sodium retention, and the animals actively seek sodium salts. One intensive study of sodium metabolism in a wild animal has focused on the moose of Isle Royale (Belovsky and Jordan 1981). Like the porcupines of the Catskills, the moose of Isle Royale face an environment impoverished in sodium. Terrestrial browse plants in the summer contain only 9 ppm (parts per million) sodium. The concentration in winter browse is even lower. Moose survive by feeding heavily on aquatic vegetation such as bladderwort (*Utricularia*), water lily (*Nuphar*), and the green alga *Spirogyra,* whose average sodium concentration of 2950 ppm approaches physiological levels in animal bodies. One of the surprising findings of the Isle Royale study has been that moose carry a great internal sodium reservoir, their rumen fluid. So much sodium is present in this fermenting stomach that the animals can draw on it during the entire year, whenever external sodium sources may not be available. Presumably a similar mechanism operates in deer, elk, cattle, and other ruminants. Evidence discussed later in this chapter suggests strongly that such an internal reservoir also exists in the caecal fluid of the porcupine.

The Salthouse

To study salt drive in the porcupine, I built my own version of a Big Bone Lick. The Catskills are a sodium-depleted region, and any salt source will attract herbivorous animals (Figure 5.1). But what source could I provide that

Figure 5.1. Schoharie Valley in the Catskills. In this mountain region, soluble sodium salts in the environment are leached out and carried to the sea. (Photo by UR)

would attract porcupines alone, not white-tailed deer, woodchuck, red and gray squirrels, and other small herbivores? I built a structure I had known only porcupines to attack—a house. And I positioned it along a known porcupine path, a route the animals followed to our orchard and to the old barn beyond. An important detail included broad eaves to keep weather off the sides. Finally I applied the attractant, in the form of salt-impregnated two-by-twos arranged as a palisade along the sides of the 10 × 14-foot cabin.

To confirm that sodium was indeed the ion the porcupines were seeking, I interspersed sodium-impregnated sticks with untreated sticks and with sticks impregnated with the salts of other metals (calcium, potassium, iron, magnesium, manganese, copper, and aluminum) and the nonmetallic ammonium ion (Figure 5.2). I was to discover later that summer's end is a poor time to study salt drive. Nevertheless, within 1 week porcupines began arriving to chew the salt sticks. It was obvious at once that only one cation was being consumed—sodium. No other salts were of interest to the animals.[1]

The next task was to determine whether the porcupines preferred a particular concentration of salt. To explore the question, I installed sets of wooden pegs impregnated with sodium chloride at three concentrations: 0.15 normal (equivalent to the concentration of sodium in vertebrate blood plasma), 0.3 N,

Figure 5.2. A male at the salthouse, gnawing on salt-impregnated sticks (light-colored). Darker sticks are unimpregnated control sticks, ignored by visiting porcupines. (Photo by UR)

and 0.45 N (equivalent to the salinity of seawater). I reversed the pegs at 2-week intervals and removed them after 30 days for weighing and replacement with freshly soaked pegs. The attractiveness of each peg was measured as the amount of wood removed per month. Typical results over 5 months in 1986 are shown in Table 5.1. For porcupines, the most concentrated salt source (at least up to 0.45 N) is the most attractive source.

Several nights spent inside the salthouse, listening as animals chewed the walls outside, gave me an insight into how porcupines explore their environment. On a July night, I arrived 15 minutes before midnight, just as a large porcupine, quills maximally erected, skulked off from the west side of the house. I did not pursue; I expected it to be back later. The night was quiet. Roarback Creek had dried up for the summer, and the late-summer chorus of crickets had not yet tuned up. I had been lying inside the cabin for less than an hour when a loud rasping began from the east wall. A porcupine had arrived. The animal followed a rhythm in its work: five or six rapid scrapes of the wood, then silence for maceration and swallowing. It had been working for only a few minutes when there was a longer silence of about 5 minutes. When the scraping resumed, it was coming from a new position, a few feet removed from the previous site. After 20 minutes, a second porcupine arrived and started chewing on the middle of the north wall. The first arrival remained at work on the east wall.

Table 5.1. Salt consumption as a function of concentration

NaCl concentration	Wood removed per peg per month (g)					
	April	May	June	July	August	Average
0.15 N	38.4	40.0	12.9	9.8	14.4	21.1
	33.8	41.9	16.0	11.2	6.0	
	35.8	25.3	13.5	14.7	3.2	
0.30 N	71.8	90.7	26.0	52.0	52.8	51.7
	50.2	34.9	40.3	28.8	13.2	
	66.9	122.9	45.7	56.6	22.5	
0.45 N	87.6	85.3	13.4	55.0	35.5	55.9
	84.6	47.0	54.8	64.7	24.9	
	87.7	52.3	35.5	61.2	49.5	
Average (by month)	61.9	60.0	28.7	39.3	24.7	

Two-way ANOVA		
NaCl concentration	$F=20.1$	$P=0.0001$
Month	$F=10.04$	$P=0.0001$
Interaction	$F=1.11$	$P=0.386$

Notes: Two-by-two pine lumber (5×5×39.5 cm) was soaked in NaCl solutions for 48 h at room temperature. After drying, the sticks were mounted along the sides of the house, interspersed randomly among 141 untreated sticks and 31 sticks treated with other salts. Sticks were removed at monthly intervals between April 4 and September 7, 1986, for drying to constant weight (48 h, 110 °C), weighing, and reimpregnation. To keep the walls covered at all times, two identically treated sets of sticks were alternated along the walls. Sticks showing excessive wood removal were replaced. Thirteen unique porcupines were captured at or near the salthouse in 1986.

I had assumed originally that each porcupine would choose a position and work there till it left. Instead, they were treating the salthouse as a sodium buffet, moving at regular intervals to sample new sticks for shorter or longer periods. At the same time, the two animals maintained a wide spacing between themselves. At 1:25 a.m., the two chewers were grinding on opposite sides of the salthouse, the first animal now on the north side, the second on the south. I felt trapped inside a bass viol as the house vibrated under their powerful teeth. This was porcupine music, the sound of the wood saw. The first animal left at 1:35, 1 hour and 5 minutes after it had arrived. The second left 10 minutes later. A third animal arrived much later, at 4:15 a.m., and chewed for only a few minutes before departing. At 5:15, in the gray of dawn, there was silence.

At other times when I have emerged from the salthouse to surprise the porcupines at work, I have found them clinging vertically, arms spread to catch the tops of adjacent salt sticks. They wait silently in the glare of my headlamp, then drop to the ground and depart. But they leave behind an interesting piece of information—the salt stick they have been chewing is wet. That means they lick their targets, monitoring sodium levels continually with their tongues. It also helps explain why no porcupines visited the salthouse on rainy nights. The rain would have leached out surface ions and made all natural salt sources unattractive.

Another question the salthouse helped answer was whether porcupines respond directly to sodium concentration or to the potassium-sodium ratio in the salt source. I impregnated pegs with increasing ratios of potassium to sodium (as the nitrate salts) and followed porcupine response (Table 5.2). The clear lesson is that porcupines do respond to the ratio of potassium to sodium, not to absolute levels of sodium. That is appropriate behavior because a high potassium-to-sodium ratio in spring vegetation may be a major cause of their sodium deficit.

Porcupines show a strong seasonality of salt use (Figure 5.3).[2] Visits to the salthouse stop almost entirely between November and March, only to resume with explosive force in April and May. The salt-feeding rate then declines through midsummer and may rise to a secondary peak in August–September.

Table 5.2. Salt consumption as a function of potassium-sodium ratio of source

Impregnating solution concentration			Wood removed per peg per month (g)				
Na	K	K/Na	May	June	July	August	Average
0.075 N	0.75 N	10	0	0	0	0	0.3
			2.6	0	0	0	
	0.75 N		0	0	0	0	0
			0	0	0	0	
0.075 N	0.075 N	1	1.4	5.0	3.0	0	3.4
			5.3	6.5	6.2	0	
	0.075 N		0	0	0	0	0
			0	0	0	0	
0.075 N		0	35.6	10.0	16.8	7.3	15.6
			25.6	12.1	17.0	0	
Average removed from sodium-impregnated sticks by month			11.8	5.6	7.2	1.2	

Two-way ANOVA		
Na ratio	$F=64.67$	$P=0.0001$
Month	$F=14.16$	$P=0.0003$
Interaction	$F=9.98$	$P=0.0004$

Notes: The same protocol was followed as in Table 5.1, except that nitrate salts were used to impregnate the wood. Sticks were randomly interspersed among 141 untreated sticks and 30 sticks impregnated with other sodium salts. Weight losses were determined at monthly intervals between May 5 and September 7, 1986.

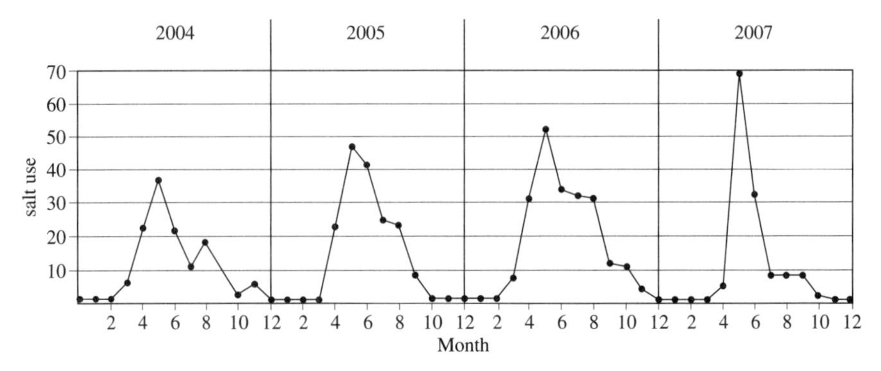

Figure 5.3. Salt use 2004–2007. The annual pattern shows a peak in May; in most years, there is a secondary peak in August.

Any explanation of salt drive in porcupines must address the strong seasonality of salt use. I return to this point later.

Another observation that needs to be addressed is the strongly skewed sex ratio of salt-chewing porcupines. The porcupines that come to cabins and outbuildings in search of salt are overwhelmingly females. Between 1982 and 2006, I was able to determine the sex of salt-chewing porcupines 344 times; 251 of the animals (73%) were females. Although the sex ratio of my porcupines in the forest is female-biased, the bias is only 58%.[3] A similar female bias in salt drive has been demonstrated in white-tailed deer (*Odocoileus virginianus*) (Wiles and Weeks 1986). Certain of my porcupine females salt-chewed so often that I thought of them as salt junkies. Rebecca salt-chewed 5 times in summer 1982 and 7 times in 1983. Squirrel did the same 5 times in 1983 and in 1991, Moth 5 times in 1984, and Loretta 5 times in 1997. In contrast, males were typically found salt-chewing only once a summer. Because salt sources were not monitored every night, all these figures are minimums. For some females, the trips to salt sources represented long journeys (up to 1 kilometer) over difficult terrain. Rebecca often made the trip in two stages. On the first night, she moved down from her rugged mountain territory and camped out in a tree near the farm outbuildings; on the second night, she moved in to gnaw salted wood, then retreated again before dawn to her normal range up the mountain. All salt-binging females had young of the year.

It is intuitively obvious why females should have a stronger salt drive than males: they lose sodium in their milk. Nursing infants do not normally visit salt sources; their mothers do. Only 3 of my 102 salt-chewing females have been juveniles (less than a year old), and all were accompanied by their mothers. But the intuitive understanding turns out to be an incomplete guide in this instance.

As long ago as 1938, Richter and Barelare showed that pregnant and lactating rats increase their salt intake, reaching a peak during lactation and dropping to normal levels soon after weaning. The phenomenon was explored in depth by D. Denton (1982) and his coworkers in Australia, who studied wild rabbits in a laboratory setting and found a tenfold increase in salt intake by the end of pregnancy, rising to a thirtyfold increase during the peak lactation period. Those spectacular rises in sodium intake cannot be accounted for on the basis of sodium used to build young bodies in utero or sodium lost in the milk. Well over 90% of the sodium taken in during pregnancy and lactation was lost in the urine. Clearly, the increased intake did not result from a sodium deficiency. Through a series of careful experiments, Denton showed that the hormones present during pregnancy and lactation, when offered in the appropriate sequence, will by themselves induce sodium drive, even in nonpregnant animals. At least four hormones are involved: estrogen, adrenocorticotropic hormone (ACTH), prolactin, and oxytocin. Denton's experiments were carried out on laboratory-confined animals exposed to unlimited amounts of sodium. They could therefore afford to lose 90% of their sodium intake. The same animals in the wild, faced with a much more limited sodium supply, would no doubt have lost little in the urine. The point of the experiment is that hormone-induced sodium hunger in the animal serves a rational need in the wild—the metabolic needs of the fetus and the nursing infant are satisfied. A behavior that is irrational in the laboratory need not be such in the field. The sometimes strange dietary cravings of pregnant humans probably have a similar basis. No comparable experiments have been performed on porcupines, but because the same hormones are present in all pregnant and lactating mammals, there seems little reason to expect a difference.

In the Catskills, baby porcupines are born in May and early June. The broad peak of salt use begins in early April. And although males make up a minority of salt-chewing porcupines, they do visit salt sources on roughly the same schedule as females. Therefore, although hormonal factors may explain the skewed sex ratio of salt-seeking female porcupines, they cannot explain the behavior entirely.

Other factors proposed as explanations for the spring salt binge in North American herbivores include changes in dietary potassium-sodium ratios and changes in dietary water content. Weeks and Kirkpatrick (1976), studying the salt-lick behavior of white-tailed deer in Indiana, show that the potassium-to-sodium ratio goes from 156:1 in winter diets to 485:1 in spring diets. The authors propose that deer kidneys, which must excrete excess potassium and conserve sodium, cannot cope with the potassium flood and allow enough sodium to pass to drive the animals into a sodium deficit. The theory is an attractive one, and it may explain the later half of the broad spring peak in porcupine salt drive. But the dietary mileposts here are feeding on spring buds and feeding on young leaves. Both food sources have higher potassium-sodium ratios than winter bark, and both follow, rather than precede, the major pulse of spring salt-feeding in porcupines.

Weeks and Kirkpatrick (1976) postulate that one route of sodium loss may involve the spring feces. Many herbivorous animals, including deer and porcupines, produce hard, dry feces for most of the year. But during spring, the feces become soft, almost diarrhetic in nature. The explanation offered is that the high dietary intake of potassium leads to high intestinal levels of potassium, which leads to decreased water resorption from the large intestine, which results in increased loss of electrolytes via the feces. I have measured sodium levels in porcupine feces by atomic absorption and found increased losses of potassium, but not sodium, during spring and summer (Table 5.3). Potassium loss peaked in May, when diarrhetic feces were produced. Fecal sodium loss did not change significantly at that time.

Another difficulty with the fecal loss theory in porcupines is the lack of correlation between salt-chewing intensity and the production of diarrhetic feces. In April, when salt drive is near maximal, the porcupines are still feeding on spring buds and occasionally on bark and producing solid droppings. The first soft feces were observed on May 13 in 1986, April 25 in 1987, and May 15 in 1988. In that respect, porcupines differ from ruminants such as the

Table 5.3. Fecal sodium and potassium loss in porcupines

Month	Sodium (mM)	Potassium (mM)
January	5.0±0	8.0±0
February	4.5±1.3	10.0±2.2
March	4.5±0.6	7.5±1.7
April	5.0±1.0	8.3±6.6
May	4.7±0.6	29.0±1.0
June	4.3±0.6	4.7±11.0
July	3.9±1.0	18.0±2.6
August	2.5±1.1	9.0±0
September	2.5±1.1	17.5±3.5
October	3.9±1.0	12.0±3.5
November	3.9±1.0	6.3±0.6
December	3.3±1.3	8.5±5.0

ANOVA	
$F_{11, 24}=6.279$	$P<0.001$

Notes: Dried feces collected in 1986 were extracted with a 10-fold excess of distilled water, and the extract was analyzed by atomic absorption. Sodium loss did not vary significantly by month (ANOVA), whereas potassium loss differed significantly by month.

white-tailed deer (Weeks and Kirkpatrick 1976) and the mountain goat (Hebert and Cowan 1971), which show increased sodium loss in spring feces. If porcupines do, in fact, suffer increased sodium loss in the spring, the route would have to be via the kidneys. I have not measured urinary sodium losses in porcupines on an annual basis.

Reproductive hormones, dietary potassium-sodium ratios, and urinary sodium loss may all play a part in the salt drive of the porcupine. One additional hypothesis bears further investigation. The near-absence of winter salt-chewing may not reflect so much a lack of salt hunger as a lack of opportunity for its satisfaction. In the Catskills, the first surge of salt-chewing correlates best with the dates for abandoning winter dens. If the animals build up a sodium deficit through a 6-month confinement in winter territory, den abandonment would be their earliest opportunity to satisfy it. The hypothesis would be strengthened if the porcupine could be shown to carry an internal sodium pool analogous to the rumen of the ruminants.

In the fall of 1986, direct evidence for a sodium pool was revealed through an unfortunate incident. An adult porcupine that I had kept captive for a feeding-choice experiment suddenly died. (The animal had been kept for 3 days on an all-apple diet.) I discovered the body while it was still warm and did an immediate autopsy, taking samples of stomach fluid as well as caecal fluid and measuring their sodium and potassium levels by atomic absorption. Both fluids contained low potassium: 9 mM in the stomach and 14 mM in the caecum. The stomach fluid contained 41 mM sodium, but the caecal fluid had 142 mM sodium. The data show that the caecal fluid is essentially a pool of isotonic saline, similar to plasma or extracellular fluid. Because the volume of the caecum is relatively large, it represents a major sodium store on which the animal can draw over the winter when normal supplies are inaccessible.

Those results have been confirmed by the work of Vispo and Hume (1995). The authors examined four adult porcupines (two males and two females) shot mid-April in an eastern white pine–mixed hardwood forest in Wisconsin. The animals were quickly weighed, then their digestive tracts removed and ligated to separate the stomach, small intestine, caecum, and colon; the latter was further subdivided into proximal and distal segments. Vispo and Hume found a sixfold increase in sodium concentration between the stomach and caecum. Although there was some spillover into the proximal colon, the sodium concentration dropped progressively along the distal colon, suggesting an efficient sodium-resorption mechanism in the large intestine.

The hypotheses examined so far have addressed the factors responsible for the spring peak in salt drive. As mentioned earlier, porcupines exhibit a secondary peak in August, which coincides with a specific porcupine behavior, the period of apple-feeding. As discussed in Chapter 6, some porcupines consume large numbers of apples in late summer, replacing this food with

Table 5.4. Correlation among apple feeding, new midsummer foliage, and salt use

Year	Apple feeding	New midsummer foliage	Secondary peak in sodium drive
2004	+	–	+
2005	+	–	+
2006	–	+	+
2007	–	–	–

available mast in the fall. For example, the 2007 apple crop failed completely because forest tent caterpillars had defoliated all the local trees in 2006, leaving the trees too exhausted to produce fruit in 2007. The salt drive for July–October 2007 was severely reduced.

But the apple crop failed in 2006 as well, and salt drive showed a strong secondary peak. In that year, most trees stood bare in June but put out a second crop of leaves by July. The large observed secondary peak may reflect the high potassium-to-sodium ratios of the regenerating leaves (Table 5.4, Figure 5.3).

Apples are an imperfect porcupine food in several respects. One shortcoming is that apples contain several hundred times more organic acid than do alternate summer foods such as linden or aspen leaves. High acid load impairs sodium resorption in the mammalian kidney (Cogan and Rector 1982; Rose 1984), causing animals to lose sodium in their urine. In fact, porcupines showed strong selectivity among apples, avoiding high-acid fruit. To test the impact of apple-feeding on porcupine sodium retention more directly, I captured a number of animals and measured their urinary sodium excretion as a function of apple versus linden-leaf diet. Table 5.5, which displays the data for

Table 5.5. Urinary sodium output as a function of the diet (for Killer)

Day	Diet	Urinary Na output (μM/h)
0	Wild	2.0
1	Red delicious apples	3.0
2	Red delicious apples	17.0
3	Linden leaves	38.0
4	Linden leaves	50.0
5	Linden leaves	0.4
6	Linden leaves	0.8
7	Linden leaves	0.8
8	Linden leaves	0.4

the adult male Killer, is illustrative. Sodium loss showed a 36-hour lag in response, increasing steadily for the 2 days the animal was on an apple diet and rising into the middle of the fourth day, when it plunged and remained very low for the duration of the experiment. The data predict that porcupines who do not feed on apples should not show a secondary peak in sodium drive. As a corollary, porcupines in sodium-rich areas such as seacoasts should show less selectivity in their apple-feeding.

Natural Salt Sources

When I built the salthouse, I designed it as a mineral supply for porcupines. Many other builders of houses and outbuildings near the woods have had no intention of seeing their houses converted to sawdust, yet porcupines visit them too. Although all large and medium-size herbivores seek salt in sodium-depleted environments, porcupines alone seem to specialize in human salt artifacts. Yet porcupines have lived in North America incomparably longer than the earliest human inhabitants. How did they satisfy their salt needs before human settlement? A look at current behavior shows that porcupines are surprisingly versatile in locating naturally occurring salt sources.

Naturalists as long ago as the nineteenth century have observed porcupines swimming to reach the yellow water lily (*Nuphar* spp.) and feed on the leaves. The porcupine swims easily, buoyed by its quills (Figure 5.4), but prefers lily pads it can reach from some solid substrate. Color Plate 5 shows a porcupine enjoying the water's edge. The great mammalogist C. Hart Merriam (1884) described the behavior a century ago:

> When feeding on lily-pads along the borders of watercourses, they sometimes utter extraordinary noises; and occasionally quarrels arise from the possession of some log which affords them easy access to the coveted plants. At Beaver Lake, in Lewis County, New York, Mr. John Constable once witnessed an encounter during which one of the contestants was tumbled into the water. The animals did not attempt to bite, but growled, and snarled, and pushed. (1884, 303)

The data of Jordan, Belovsky, and coworkers show why water lilies are such a prize. The yellow pond lily is the richest sodium source among all plants surveyed (Jordan et al. 1973); its sodium content is 9375 ppm, as opposed to an average level of 9 ppm in terrestrial vegetation. That is why moose seek these plants and others below the water's surface. Other high-sodium aquatics consumed by the porcupine include arrowhead leaf (*Sagittaria*) and the aquatic liverworts, as described by Curtis and Kozicky in Maine: "One animal was observed wading in a shallow pool, with water up to its belly. It kept its tail

Figure 5.4. Porcupine swimming. Its quills add to buoyancy. (Photo by UR)

erect. It was found to be feeding on aquatic liverwort (*Riccia fluitans*), which it devoured with relish and in some quantity before detecting the observers" (1944).

The Vly Mountain area in the Catskills is poorly supplied with pond vegetation. In early September 1982, I discovered a porcupine using an alternate salt source. During a round-the-clock monitoring of the female Squirrel, who was resting in an old sugar maple, my student Christine encountered a bobcat. It came down silently from the mountain, stopped 12 feet from my student, and stared at her. After what seemed like a long silence, it turned and moved back up the hill. Christine said her heart was pounding. That night, we discovered that the bobcat and the porcupine may have been interacting in a roundabout way. My second student, Cathy, took the dusk-to-midnight shift, when our porcupine should descend from her rest tree and start feeding. Promptly at 9 p.m., Squirrel descended from the rest tree, moved about 40 feet uphill, then began rasping at a hard object with her teeth. She rasped away for 20 minutes until Cathy, overcome by curiosity, turned on her light and moved forward to investigate. Squirrel scooted up a nearby ash tree for refuge. Perhaps because she was looking for a tree, Cathy could not find the object Squirrel was gnawing on. The next day we retraced Squirrel's route and found the object almost at once—the front and hind leg bones of a deer. The vertebral column and ribs were a short distance away. The deer-kill was relatively fresh, the bones free of flesh but retaining shreds of tendons, ligaments, and skin. It was clearly what the bobcat had come to pick at. But what a chewing the porcupine had given the remains! Every long bone was scraped as though it were made of soft plastic, exposing the marrow and sometimes leaving only a thin spiral of hard matrix. I studied the surfaces carefully and saw everywhere only the large, powerful groovings of porcupine teeth. There was no evidence of the smaller teeth of mice at work.

The primary mineral constituent of bone is calcium phosphate. Calcium is so abundant in the normal leaf diet of porcupines that it is excreted in high concentration in the urine.[4] Evidence from the salthouse shows that porcupines make no effort to seek that mineral. Fresh bone, however, also contains a significant amount of sodium. Some 25% of an animal's sodium pool is tied up in a poorly exchangeable form in its bones (Robbins 1983). That seemed to explain Squirrel's interest in the deerkill. Walter P. Taylor (1935) also describes a single instance of deliberate scavenging. He quotes M. E. Musgrave, an employee of the Biological Survey, who observed a porcupine in the open desert near Bonito, Arizona. The animal was gnawing the dry carcass of a cow.

That brings up a related point. Although porcupines are commonly known as strict vegetarians, Taylor, in dissecting many porcupine stomachs in Arizona, notes as a curiosity that three stomachs contained animal matter: one held bits of a beetle plus an ant, and two others held ant fragments. Taylor assumes they were ingested accidentally. In collecting fresh linden leaves, I have often found caterpillars clinging to the edge or hidden inside a rolled leaf. Common varieties in the Catskills include the larvae of the fall cankerworm (*Alsophila pometaria*), the fall webworm (*Hyphantria cunea*), the linden looper (*Erannis tiliaria*), and the American daggerwing (*Acronicta americana*). A porcupine feeding in total darkness may inadvertently ingest those tiny packages of sodium (Figure 5.5). But apparently that source alone cannot satisfy the animal's sodium needs because the majority, at shorter or longer intervals, continue to seek additional sodium supplies.

Seastedt and Crossley (1981) point out that the outer bark of trees is surprisingly rich in sodium, averaging 546 ppm, probably deposited when stemflow is followed by evaporation of water. Such a source should be most rewarding in dry environments. The ever-observant Walter Taylor (1935) notes that porcupine may on occasion feed on outer bark, a behavior that must be clearly distinguished from winter bark-feeding, when the inner portion of the bark is consumed.

Porcupines are also able to discover and harvest a novel sodium source in the forest. The New York state Department of Environmental Conservation maintains a network of trails through the Catskill State Forest. In years past, circular aluminum trail markers—about 7 cm in diameter and painted red, blue, or yellow—identified the different trails. The markers were nailed to

Figure 5.5. A possible natural sodium supplement—caterpillars of the fall webworm (*Hyphantria cunea*) on underside of linden leaves. (Photo by UR)

trees along the trail, spaced about 50 feet apart. While climbing from Devil's Tombstone to the top of Plateau Mountain, I encountered a long string of trail markers with the paint rasped off in a sunburst pattern (Figure 5.6). Bare aluminum gleamed from the markers. The long tracks of the scrapes identified the worker as a porcupine. The animal (or several animals) had climbed each trailside tree in turn to reach the marker, then held on in impossible positions to scrape the paint. None of the trees I examined has been otherwise touched; there were no niptwigs and no signs of bark feeding.

To establish that the animals did not have an unexpected need for aluminum, I nailed 53 clear aluminum markers to trees along one of our own trails. None has been disturbed in years of observation, despite heavy porcupine traffic along the trail and leaf-feeding in some of the marked trees. The Plateau Mountain animals are therefore hunting the sodium in the blue paint (whose presence I determined by atomic absorption measurement). Their chewing is remarkable because each marker represents only a tiny packet of sodium and because it must be a learned behavior developed in a systematic way. The porcupines know their forests extremely well. New York state, incidentally, has now switched to plastic trail markers.

Figure 5.6. A novel sodium source—blue paint on a trailmarker. (Photo by UR)

The salt-seeking described here makes the porcupine, like the reindeer, a proto-domesticated animal. Salt is a commodity the porcupine needs so badly that it will seek it from human sources. In so doing, it risks death and injury in challenging the most powerful predator in its environment—human beings.

Notes

1. Two-by-two pine lumber (5×5×39.5 cm) was impregnated with 0.15 N aqueous solutions of the sodium salts NaAc, Na_3 citrate, $Na_2B_4O_7$, Na_2CO_3, NaH_2PO_4, and Na_2SO_4, and with the nonsodium salts $AlCl_3$, $AlNH_4(SO_4)_2$, $Al_2(SO_4)_3$, $CaCl_2$, $FeSO_4$, K_2CO_3, K_3 citrate, KCl, K_2HPO_4, KH_2PO_4, K tartrate, $MgCl_2$, $MnCl_2$, NH_4Cl, $NH_4H_2PO_4$, and $(NH_4)_2SO_4$. All salt sticks were prepared in duplicate and mounted randomly among 133 untreated sticks of the same size (ca. 250 g each) and shape. The salthouse was erected July 30, 1984. Between August 18 and September 23 no nonsodium sticks showed signs of chewing; all Na-containing sticks were chewed, with an average weight loss of 8.77 g ± 8.86 g per stick. Between August 3, 1984, and the end of the summer, five unique porcupines were captured at or near the salthouse.

2. Salt drive was expressed as weight loss per stick per 30 days. Wood was impregnated with 0.3 N NaCl and mounted as described in note 1. Fifteen unique porcupines were captured at the salthouse in 1987 and the same number in 1988. Wood chewed per porcupine, March

13–August 1, was 485 g in 1987 and 471 g in 1988. For comparison, eight unique porcupines were captured at the salthouse in 2007, and the wood chewed per porcupine was 495 g.

3. Of salt-chewers between 1982 and 2007 (including repeat visits), 93 were males and 251 females. Of unique animals captured over the same period, 74 were males and 117 females. $\chi^2 = 7.82$, df = 1, $P < 0.01$.

4. Atomic absorption cation values for urine obtained by catheterization of an adult female are Na 0.6 mM, K 106.9 mM, and Ca 101.8 mM.

6 Summer and Fall Foraging

The active life of the porcupine, like that of most forest mammals, begins at dusk. Animals that may have spent the whole day in comfortable sleep on a tree branch descend in a darkened forest, travel to a feeding tree, and climb into the canopy for the nightly meal. By staying up the night with a foraging porcupine, one may encounter other forest mammals, as well as owls, giant millipedes, and legions of moths, midges, and leaf-feeding caterpillars. And one may discover that the night forest is noisier and more full of life than its daytime counterpart.

A Night in the Woods

On a July night in 1982, I set out to observe Rebecca, a young adult female captured and collared only a week before. Earlier in the day, I had tracked her to a linden tree on a steep hillside. As I move after nightfall, her radio signal pulls me back in the same direction. I am laden with gear: a miner's headlamp, radio receiver and directional antenna, rolled foam mattress with groundsheet, rucksack with extra jacket and poncho, altimeter, a small tape recorder, and gear for capturing porcupines. I reach the animal at 10:15 p.m. At first I don't recognize the surroundings in the dark. Then, 15 feet from Rebecca's feeding site, the aroma of porcupine urine envelops me. The ground downhill from the tree is saturated with it, but when I reach the main trunk, I smell nothing. Although porcupines seldom sleep by day in the same trees they feed in at night, Rebecca is up the same large linden tree I left her in that afternoon. She is too high to see, and the antenna does not discriminate with pinpoint accuracy. But proof comes at 10:21, with the sound of a large branch falling. That is followed by a porcupine dropping rattling down at 10:25, then more at 10:30, 10:36, and 10:42. The fallen branch is a classic niptwig, about 18 inches long, with the base cut cleanly at a 30-degree angle and leaves consumed down to the petioles. It carries some immature linden fruit—rejected along with the petioles. Rebecca is feeding highly selectively in total darkness.

On a level spot about 12 feet uphill from the trunk, I unroll my mattress and groundsheet and lie down to listen in the dark. A distant bird cries out sleepily. The porcupine rustles in the canopy overhead. A fine patter of caterpillar frass is falling—caterpillars, like porcupines, feed mostly at night and seem especially active in this part of the forest. The quiet sounds are punctuated by the thumps of porcupine droppings hitting the dry leaf litter, falling at the rate of approximately 15 per hour, sometimes in barrages of 5 or 6 at a time. The droppings are the size of deer pellets, with a gently crescent shape. At this time of year, they are almost dry.

Deer mice keep up long bouts of squeaking, so high-pitched they border on the inaudible. Periodically, some beetle buzzes against the dry leaves to get airborne. Most surprising is the heavy foot traffic along a small lumber road, perhaps 50 feet downhill, out of range of my flashlight. All night long I hear animals heading uphill and down. Some come close, catch my scent, and retreat. Others complete the passage and grow inaudible in the distance. Many are probably deer, but some may be porcupines, and I wonder whether Rebecca followed the same path to her present feeding site. Not all the passages are innocent. At midnight, I hear a small death scream in the distance. At 1:15 a.m., there comes the heart-stopping, choking-man cry of a black bear.

Despite the traffic below me, Rebecca dominates the night. In addition to the irregular bombardment of droppings, I hear large-scale movements, as though a branch is being shaken. Sometimes there is a grunting cough. At 1 a.m., a shower of urine splatters down; all of it seems to be caught by the understory of striped maple above me. A second shower comes at 4:25, as the sky begins to lighten. Above all, I hear the insistent tearing and crunching of linden leaves as Rebecca feeds in almost a frenzy in the early part of the night, to about 1 a.m. According to my tape recorder, the feeding tapers off in the small hours of the morning, but I doze off several times in the cocoon of my poncho.

At 6:15, a deep orange sun hangs in the green undercanopy, making it glow. The birds have been singing a loud morning chorus for an hour already, drowning out the subtle rustlings of the porcupine above. A wood thrush sings close to my position while remaining hidden from view. I stand up. A moment later, a small-headed fawn comes bounding down the hillside. Its legs seem to belong to some larger animal; its trunk is still dappled. Some 20 feet away, the fawn sees me and stops, moving its head up and down, plainly in a conflict between fright and curiosity. Suddenly the wind is right and it catches my scent. Launched by a jump, it floats down the hill on its long legs.

Under Rebecca's tree, droppings continue to fall at the same rate as earlier in the night. At 7:19, a half-chewed fragment of linden leaf drifts down. Rebecca is still feeding! I collect my gear and stumble home, thankful to be moving downhill.

My experience with Rebecca was atypical in one sense. As mentioned before, porcupines don't normally rest in the same tree by day that they feed in at night. Apparently Rebecca's feeding tree was large and comfortable enough to serve as a resting tree as well. She returned to it several times during the summer.

Why do porcupines feed at night? An obvious consideration is predator avoidance, although most of the porcupine's natural predators are also active at night. Fournier and Thomas, in a 1999 study of porcupine winter thermoregulation, found the animals' core temperatures rising an average of 1° C at night. Whether the same is true in the summer remains to be determined. A final possibility that deserves experimental evaluation is a change in plant chemistry from day to night. Tree leaves become saturated with water and may undergo changes in acid-base balance. Whatever the reason for the porcupine's activity preference, it introduced me to the rich life of the summer night.

Linden

Before tracking porcupines through the night, I had been only dimly aware of the linden (*Tilia americana*) as a Catskills forest tree. It is rare along the slopes of Vly Mountain; fewer than one tree out of a hundred is a linden. In addition, the nondescript bark resembles ash, and the distinctive leaves and fruits are often so high up in the canopy they can be seen only with binoculars. After two summers of radiotracking, I came to realize how much the porcupine's summer life revolves around this tree. In a curious way, the linden began to peer out of the crevices of the forest. One summer morning, during a hike through a distant cove, I stopped in surprise when a sweet fragrance enveloped me. I had caught the bee's message—a linden tree in full blossom. In the Catskills forest, almost all the tree species (and all the most abundant species) are wind-pollinated. The linden is visited by bees, who can find trees widely scattered in the forest. The linden's ability to keep in genetic touch at a distance helps it to persist in the forest at a low density.

But that low density proves no barrier to the tree's discovery by porcupines. Often, as I followed a porcupine at night from resting tree to feeding tree, or from one feeding tree to another, the tree at the end of the journey proved to be a linden. From midsummer on, when beech (*Fagus grandifolia*) and white ash (*Fraxinus americana*) dropped out of the diet, the linden became the single most important food species. During the summers of 1983 and 1984, linden leaves were consumed in 34% of all feeding episodes. Other foods included apples (30%) and two aspen species (big-tooth and quaking), which made up 26% of the total. The remaining 10% of feeding episodes made use of yellow birch, shadbush, American elm, and raspberry canes (Figures 6.1 and 6.2). The remarkable feature of the porcupine's summer diet is that the food species, with the exception of raspberry canes, are rare or absent in the forest (Table 6.1). The animals must leave the forest to feed on species such as apple.

Linden leaves were consumed in the same manner as beech and ash were earlier in the year: small leafy branches were nipped off, the leaf blades removed, and the branch with petioles discarded as a niptwig. Midsummer nitrogen analyses showed that porcupines select only the most nutritious part of the tree, the leaf blade, which contains 13.1% crude protein. The discarded parts (the petioles and twig tips) contain only 3.5% and 4.6% crude protein, respectively.

Whenever I homed in on an animal's radio signal and finally discovered where it was feeding, I measured the feeding tree and labeled it with a small plastic marker. All linden trees tend to look alike at night in the beam of a headlamp, but as I walked around the trunk, marking and measuring, I often

Table 6.1. Relative densities of porcupine summer food trees

Tree species	Use (%)	Relative density in forest (%)
Linden (*Tilia americana*) leaves	34	0.9
Apple (*Pyrus malus*) fruit[a]	30	0.0
Aspen (*Populus grandidentata, P. tremuloides*) leaves[a]	26	0.1
Yellow birch (*Betula alleghenensis*) leaves	2	2.0
Shadbush (*Amelanchier canadensis*) leaves	4	0.6
American elm (*Ulmus americanus*) leaves[a]	1	0.0
Raspberry (*Rubus* spp.) canes	3	N.A.

Notes: Relative density = (Number of stems of species *i*/Total number of stems) × 100; N.A., not applicable

[a] Forest-edge species

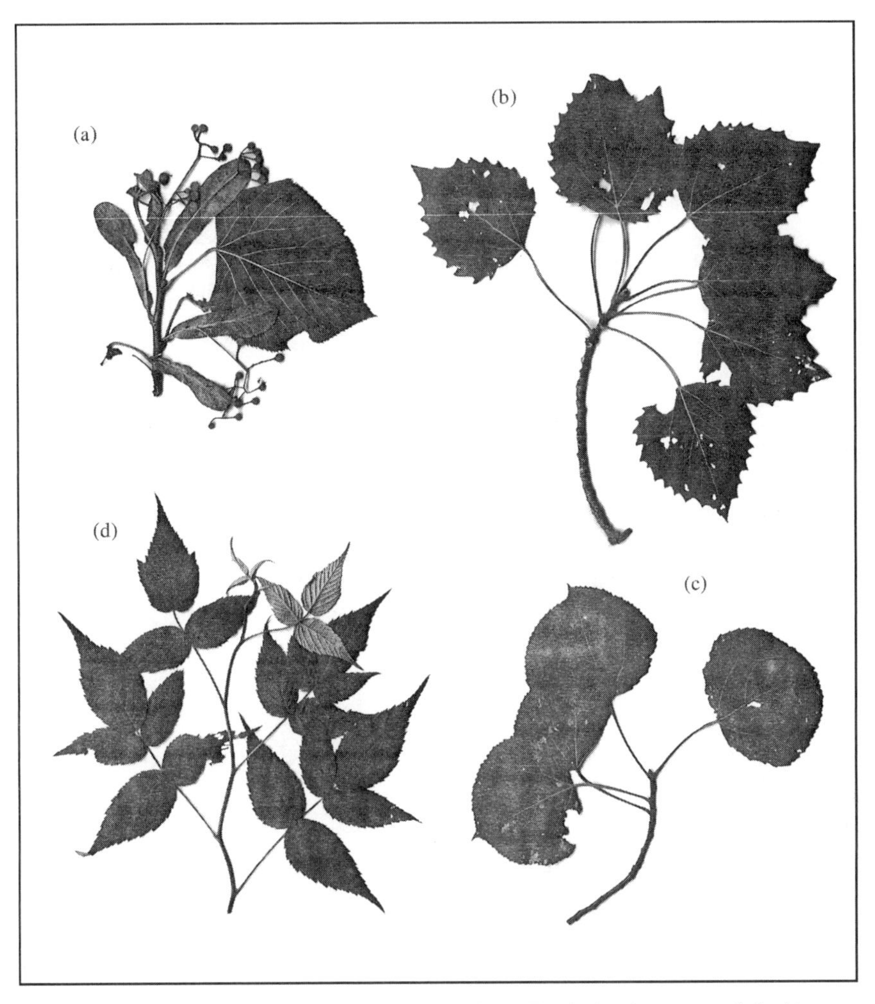

Figure 6.1. Major leafy summer foods of porcupines. Clockwise from upper left: (a) Linden (with inedible fruit) (b) Bigtooth aspen (c) Quaking aspen (d) raspberry.

spotted the gleam of a small plastic marker already fastened to the tree. The porcupines were returning to trees they had visited before. Although lindens are rare in the Catskills forest, occasionally a feeding tree was close to another linden that seemed to attract no porcupines at any time. It appeared that porcupines were discriminating at three levels: the level of tree species, the level of individual tree within the species, and the level of plant part in the tree (leaf blade vs. petiole). Nitrogen measurements could explain plant part preference—leaf blades contained more crude protein than any other part of the plant. But I could not explain the preference for individual trees. I decided

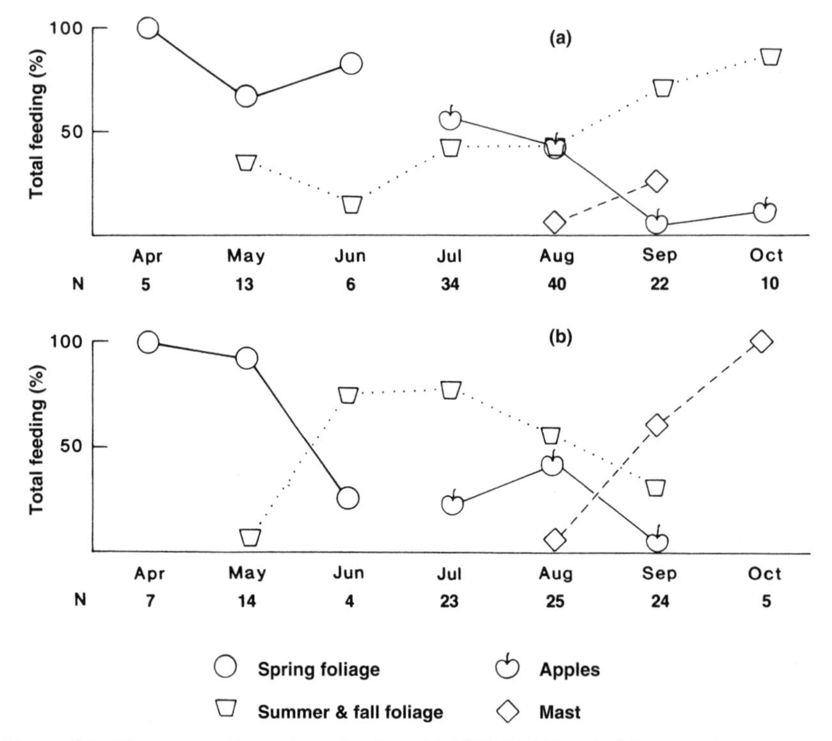

Figure 6.2. Two years of nonwinter feeding. (a) 1983 (b) 1984. A rich acorn (mast) crop in the second year shortened the period of apple feeding.

to monitor several linden tree regularly to measure the variation in feeding pressure, then analyze the preferred trees chemically to see what made them attractive.

During summer 1984, I made weekly visits to 13 widely scattered trees to measure the extent of porcupine feeding in each tree. The group was expanded to 17 trees during 1985. Niptwig count became my measure of feeding intensity. By carefully searching the ground under each tree, I could recover even a single niptwig. The green petioles on the branches served as visual flags against the brown forest floor. In both years, the tempo of feeding started slowly in June, then quickened as the summer developed. Peak feeding intensity was reached in early October, just before leafdrop. Lindens and aspens were among the last tree species in the forest to hold on to green leaves, after the dominant maples and beech stood bare.

The porcupines' selectivity was startling. During summer 1984, I found 151 niptwigs under the most preferred tree and 58 under the second most preferred. The remaining 11 trees combined contributed only 25 niptwigs; 8 of them were never used at all, though some grew relatively close to preferred trees. The two most preferred trees accounted for 90% of all linden feeding within the cohort. In 1985, the pattern of unequal feeding preferences continued, with less dramatic inequalities. Porcupines may have used some less-preferred trees because total feeding pressure was greater. (The local porcupine population increased in 1985; see Chap. 10.) Nevertheless, four trees still accounted for 69% of all feeding (Table 6.2).

My next problem was how to collect leaf samples for a variety of chemical tests. I designed my own low-tech leaf collector, a pair of three-quarter-inch steel washers tied together with twine and shot from a powerful slingshot. Every week, I became a hunter of leaves. A long pull on the slingshot, the washers flew up with a whining sound, and a small thwack of contact in the canopy, For a few moments nothing happened; then I saw my game, a large leaf sailing down in reversing arcs. My eyes never left it until it was on the ground. I bagged it.

Table 6.2. Intraspecific selection of lindens, 1984 and 1985

	Total niptwigs (%)	
Tree	1984[a]	1985[b]
A	4	4
A′	65	10
B	0	0
B′	1	1
C	0	0
C′	0	0
D	0	7
D′	5	10
E	0	1
E′	0	2
F	—	2
G	—	8
H	—	5
J	0	0
K	0	2
766	—	19
853	25	30

Note: Trees within 100 feet of each other are identified by a common letter (e.g., A and A′).

[a] In 1984, Total number of niptwigs counted=234; Average niptwigs per tree=18

[b] In 1985, Total number of niptwigs=625; Average niptwigs per tree=36.8

Shooting leaves on an August afternoon, I was nearing the end of my circuit and trying to collect leaves from D', a large, sloping tree on a woodland trail. The sun had just set behind the mountain, and the trail lay in shadow, but the tops of the tallest trees still caught the last rays. I launched my projectile and heard it rise through the canopy with only a tiny tick—no leaf flew down. But as the bright washers kept soaring, they entered the zone of high sunlight and twisted, luminous and free, before arcing back to the shadows. As the missile tumbled and fell, I had an illumination. I realized that the porcupine, too, was a hunter of leaves. Unlike the grazing ruminants of the plains that crop a field of grass and find any number of species acceptable fodder, the porcupine must hunt its food by discovering the rare trees in the forest that will best sustain it. Out of a thousand forest trees in the Catskills, one or two are acceptable lindens and one is a big-tooth aspen. To locate and return to those trees, the porcupine uses all the skills of the hunter: memory, attention to landmark and local detail, persistence. That explains why the American psychologist L. W. Sackett (1913) found that the North American porcupine has an extraordinary ability to learn complex mazes and to remember them as much as a hundred days afterward.

But I was hunting for more than leaves—I was also hunting for understanding. No single factor had leapt out as an explanation for the porcupine's choices, so I examined many possibilities. Were the preferred trees richer in nitrogen, in the same way that leaf blades are richer than petioles? I wasted much effort on those onerous analyses because the nitrogen values showed a great deal of scatter. Individual leaves from the same tree may differ by several percentage points of crude protein. But, in the end, a statistical analysis of results showed no correlation between nitrogen content and individual-tree feeding rank. In fact, the average nitrogen content of the trees fell during the course of the summer as overall feeding intensity rose. Are nonselected trees better protected by tannins? I ran two kinds of tannin determinations (condensed tannins and total phenols) and found enormous variation among trees. But, again, no measurement correlated with feeding preference. What about levels of calcium, sodium, and potassium? Porcupines have a strong salt drive. Perhaps they were responding to subtle differences in sodium or potassium levels. That proved to be another dead end.

Finally, in early October 1984, I looked at the acid-base balance of linden leaves. I had found earlier in the summer that porcupines, when feeding in apple trees, avoided trees with acid (low-pH) apples. I collected leaves from each tree in the group, homogenized a leaf-water mixture in a blender, and measured the pH with a portable pH meter. For the first samples, the pH values showed little variability, falling in a mildly acid range between 5.9 and 6.2. All those came from low-ranked trees. Then a leaf from tree 853, the porcupines' second-favorite tree in 1984, registered a pH of 6.7, nearly neutral. I quickly homogenized leaves from tree A', the animals' first choice, and got a value almost as high. Finally, a pattern was discernible (Figure 6.3).

My 1984 results were based on a single set of pH readings taken at the end of the feeding season. I repeated the pH measurements in 1985, taking readings throughout the summer. The results not only confirmed that porcupines feed most heavily on trees with highest-pH leaves but also that the overall tempo of feeding parallels a seasonal increase in pH.[1] Taking the pH readings

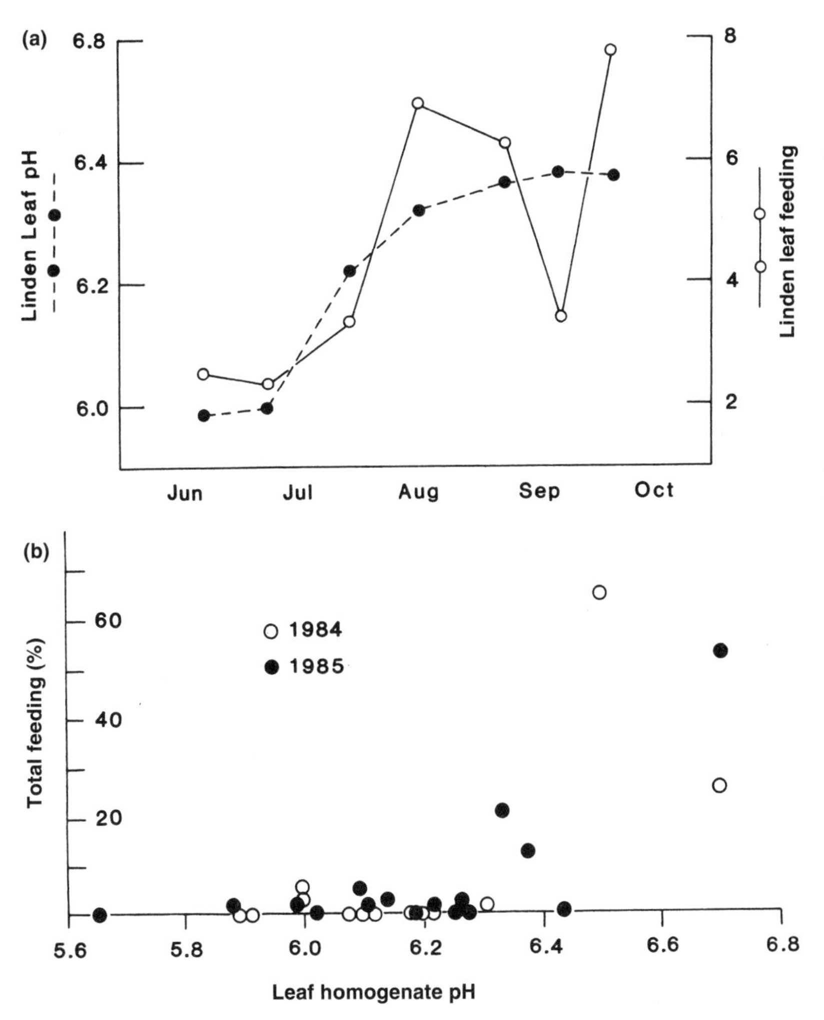

Figure 6.3. Linden consumption and pH. (a) Time course of linden use and average leaf pH in 1985 (b) Relationship between feeding pressure on individual lindens and pH of their leaf homogenates.

at face value, we might assume that porcupines want to limit acid intake, as they undoubtedly do in the case of apples. But the apples that they consume in great quantities show acid contents two orders of magnitude higher than in the linden leaves. It is therefore more likely that porcupines are responding not to acid content per se but to some associated defense compound in the leaves. Unfortunately, the nature of the compound is unknown at this time. Whatever the acid status of linden leaves may be, they do not offend human taste buds. I have eaten linden leaves from both low-pH and high-pH trees and found them indistinguishable. They were surprisingly mild, almost bland in taste. In that respect, lindens are very different from the impossibly bitter aspen leaves, which also rate high on the porcupine summer menu.

Aspen

The only trees challenging the linden in importance in the summer diet of the porcupine are bigtooth aspen (*Populus grandidentata*) and quaking aspen (*P. tremuloides*). Aspens are even more scarce in the Catskills forest than the lindens. Their aggregate abundance in the forest is approximately 0.1% (i.e., one tree out of a thousand is an aspen). Yet 26% of the animals' summer feeding was on the leaves of these two species.

Aspens are among the most shade-intolerant of all forest species (Bakuzis and Kurmis 1978). Their seedlings cannot survive in the deep shade of beech, maple, ash, and red oak. Although bigtooth aspens can reach full canopy height, they do not replace themselves in the mature forest. Quaking aspens grow only along the forest edges, and porcupines at times undertake long trips to reach them.

In my observation, porcupines showed none of the preference for individual aspens that they show for lindens. They do harvest leaves using the same niptwigging technique and avoid the petioles. As with linden, aspen leaf blades have much higher nitrogen levels than petioles or twig tips.

A different conclusion was reached by Diner and colleagues (2009), who studied porcupine use of quaking aspens in Quebec. They demonstrated that porcupines do feed selectively on aspens, not at the level of individual trees (ramets) but at the level of clones (genets). Aspens propagate vegetatively by root sprouting, forming genetically identical clones. Their clones differed little in leaf nitrogen, starch, or sugar content; however, the levels of two phenolic glycosides, tremulacin and salicortin, varied by a factor of up to 3.9 between clones. They find that it was the levels of phenolic glycosides that dictated feeding preference, with porcupines avoiding the high-glycoside clones.

Aspens present a danger encountered in few other tree species—brittle branches. One night I tracked Rebecca to a young quaking aspen growing on the edge of Roarback Field. After waiting a few minutes for her to resume feeding, I continued up the hill to visit my other animals. Next morning, on the ground under the aspen, I saw a branch about 4 feet long, with little sign of leaf removal. The base showed a jagged, splintering break, not the clean, diagonal cut expected of a niptwig. I was holding the leader of the tree. Rebecca had climbed into the top of the tree and bent it past its breaking point. Since I found no quill mass on the ground, she must have saved herself by snagging another branch as she fell. Her free movements in the days following also suggested no injury.

A different fate awaited Crew, Squirrel's son of 1997. In the fifth month of his life, he fell to his death from a tall bigtooth aspen. His story is further explored in Chapter 14. Hale and Fuller (1996), working in central Massachusetts, report similar observations. Of 16 porcupines who died over the course of a year, two fell to their deaths out of bigtooth aspens while feeding in the brittle branches.

Apples

Apple trees are Old World immigrants in North America, not a normal component of the native forest. Nevertheless, wherever orchards lie close to porcupine-inhabited forests, they become porcupine magnets. Porcupines love apples at any time of the year. One winter afternoon when I wanted to show a porcupine to two visitors, we made our way up the mountain, through deep snow, to Finder's den, a low rock outcrop just below the ridge. The porcupine was deep inside the earth and out of sight. I placed two Red Delicious apples in the den entrance, then stepped back to watch. We stood about 12 minutes in the deep silence of a snowy landscape. Then a little movement commenced inside the den. Finder's head emerged from the shadows, picked up an apple, and dragged it inside. We heard the soft sounds of a porcupine enjoying its favorite food. Even captive wild porcupines, who normally turn their backs and bristle all quills at human approach, will turn and take an apple out of my hand.

These apple fiends begin to raid our orchard by early July (Figure 6.4). The orchard itself, with 16 apple trees and one pear, covers 2 acres. To the north and west are old fields reverting to woods and containing a dozen-odd wild apple trees derived from the orchard seed source. With a single exception, the apples of greatest interest to my porcupines have proved to be not the orchard fruit but the wild trees. Raccoons like them too. On a bright-starred, moonless

Figure 6.4. The orchard. Every year, the apple tree on the right of the footpath was stripped of its apples by mid-August, but surrounding trees stood untouched until fall. (Photo by UR)

night in July, I tracked Squirrel to a wild apple tree in Roarback Field, northwest of the orchard. There was a suspicious amount of movement in the tree, and I turned on my head lantern. Clinging to the branch above Squirrel was a baby raccoon, green eyes gleaming in the light. The two animals were sharing the tree without interfering with one another. In years since, I have found raccoons and porcupines sharing the same apple tree many times, without a sign of discord. Squirrel moved energetically through the tree, harvesting apples directly or cutting small niptwigs to gather fruit on short terminal branches. She ate only the apples and had no interest in leaves. In the beam of my lantern, I sometimes caught the dull red reflections of her eyes, sometimes the white top of her head, sometimes a reflection from her radio box. For the most part, she remained hidden behind leaves, and I saw only the branches moving as her feet sought separate footholds for stability.

Later in the month, Squirrel started bringing her baby to feed in the apple trees. Mother and baby foraged in the same tree but on different branches. The baby was relatively clumsy; during one 40-minute observation period, it dropped 18 apples to the ground. Squirrel almost never dropped her food. She turned the apple carefully in her paws, biting off small pieces till only the core (endocarp) remained. In the morning, the discarded cores displayed her careful workmanship—only the soft outer meat (exocarp) had been removed (Figure 6.5). Presumably porcupines reject the cores to avoid ingesting the cyanide-rich seeds.[2]

Many porcupines came to feed on apples. I counted at least eight different animals during a summer, including two different mothers with their babies and two adult males. It was not unusual to find two animals feeding in the same tree, and on two occasions I found three. They tended to crowd into a single tree because only a small number of the apple trees were acceptable to the porcupines: one tree in the orchard proper and six wild trees on the periphery. The difference between acceptable and rejected apples was obvious from a simple taste test; the rejected apples were sour. Preferred apples were also the first to mature.

In chemical terms, the selected apples had an average pH of 4.07, the rejected fruit averaged pH 2.95. The two classes of apples did not differ significantly in other respects. They had comparable levels of total sugars, crude protein, and sodium-potassium ratios. Porcupines have good reason to limit

Figure 6.5. Adult porcupine consumes an apple. (Photo by UR)

their acid intake. Excess acid reduces the efficiency of physiological sodium conservation (Cogan and Rector 1982; Rose 1984). As noted in the previous chapter, porcupines feeding on apples showed increased sodium drive and lost sodium in their urine. Presumably such losses would have been greater if the animals had not selected the least acidic fruit. To humans, the preferred apples had a bland, mealy taste. To porcupines, they were so attractive that preferred trees were stripped of their fruit by mid-August. The trees stood with a rumpled look, their branches littered with wilted niptwigs. Nearby, trees with more-acidic fruit stood untouched. In years when the mast crop failed, porcupines returned to the less desirable trees in September as the apples matured.

Dodge (1967) has commented on the porcupine's special preference for Red Delicious apples. I bought four strains of apples from a greengrocer and measured the pHs of their homogenates (1:10 in water). The results might have been predicted from the acid-avoidance theory. Red Delicious had the highest pH, at 4.1. It was followed by Yellow Delicious at 3.7, Macintosh at 3.35, and Granny Smith, the most acidic, at 3.1. Red Delicious apples do not differ significantly from other strains in terms of protein content (0.6% for Red Delicious, 0.7% for Macintosh). Raccoons, who are less efficient apple foragers, did not follow the same apple preferences as porcupines.

During the peak apple-feeding months of July and August, porcupines never stopped consuming tree leaves. The reason may lie in the low nitrogen content of apples (about 0.7% crude protein for apples vs. 15% for linden or aspen leaves). In fact, porcupines who fed on apples always fed on linden or aspen leaves during the same night. During 12 all-night observation sessions, animals feeding on tree leaves alone used an average of 1.25 trees per night (range: 1–2). Five animals feeding on mast or apple fruit used an average of 3.2 trees per night (range: 2–5), always accompanying mast or fruit with tree leaves.[3] The apples' high content of digestible carbohydrates, however, provided an energy source and allowed the porcupines to continue putting on weight during the summer.

Fall Foraging

At the end of August, as the favored apple trees stand depleted, the forest yields a major new food source—mast in the form of acorns and beechnuts. Even though the porcupines have not climbed beech trees since spring and have climbed oak trees only for occasional daytime rest, the two tree species now assume overwhelming importance for the porcupines, particularly in years of heavy mast crops.

Red oaks and beechnuts produce a small, basal crop of acorns and beechnuts every year. But in certain years, all trees in an area fruit massively, and acorns sufficient to establish several forests lie on the ground. Such overproduction is necessary to outstrip the demand of acorn predators: the weevils, chipmunks, gray and red squirrels, white-tailed deer, wild turkeys, raccoons, and porcupines. All descend on the crop with such force that in basal years no acorns may survive. For red oak, 1983, 1986, and 1988 were basal years; 1984, 1985, and 1987 were years of heavy mast crops. Porcupines were able to take a good percentage of the crop in each year because they got to the acorns before the ground predators did.

Again, porcupines used the niptwigging technique to harvest acorns. (Beechnuts, producing a relatively minor crop every year, are harvested in the same manner as acorns.) The porcupines cut small terminal branches, then scoop the acorns out of the acorn cups or the beechnuts out of the beechnut husks. The residue is discarded. The canopies of large oaks quickly fill up with snagged niptwigs. The giant trees hold enough fruit to keep a porcupine in residence for several days. Smaller trees, especially in basal years, might be cleaned out in minutes or hours.

One acorn-foraging porcupine that I followed through an entire night in late September climbed six different trees, three of them oaks. Mayday, an adult male, had spent the day resting in a large sugar maple. When I reached him at 9:50 p.m., the warmth of the day still hung on the mountain, but Mayday was already crunching leaves in his first feeding tree, a small linden. Acorns are an excellent energy source because of their high fat content, but like apples, they are poor nitrogen sources. That is why porcupines continue feeding on tree leaves throughout their mast harvest, up until leaf-fall. Mayday's feeding had become more frenzied by about 11 p.m.; I heard branches dropping and the animal moving in the canopy. There was a rich smell of fallen leaves and moisture.

At 12:08 a.m., an owl screamed out from a tree so close that I felt as though I were sitting inside the horn of a giant loudspeaker, with vibrations rubbing my skin. As I stared into the forest, trying to locate the source of the sound, a large, dark bird landed in the tree behind the porcupine. Its wings were outlined in flight against the transparent canopy, the sky alight with stars. Mayday had grown very still. The owl soon left on silent wings, hitting no tree branches, as would a wild turkey or grouse. When it reached its new perch, it screamed a single note, with a twist and crackle in the middle, like the first note of a siren. The porcupine did no more feeding in his linden tree.

At 2 a.m., Mayday descended and moved off into the night. I gave him a head start, then followed his radio signal upslope along the ridge. The head start I offered had not been enough. Half a mile farther and 200 feet higher up the mountain, I found the animal in a small red maple, a species never used as food by porcupines. Mayday had climbed the tree to escape me. Again, I waited in the dark. Once Mayday descended to a low crotch in the tree and chattered his teeth at me. Then his courage failed, and he climbed back to the top. Finally, 20 minutes after my arrival, he left the tree. I gave him a 10-minute head start. When I caught up at 3:40 a.m., Mayday had moved a short distance south while remaining at the same elevation. He was up another red maple, his third tree of the night. The tree was good-sized. With all its leaves gone, it appeared hung with stars. I lay down near the tree and watched.

I wondered why he had chosen a tree with no available food. All around us, the area was thick with oaks, the ground littered with niptwigs and empty acorn cups. The answer came quickly. The oaks were already occupied. In one oak, a porcupine was feeding noisily. From another oak came the sound of droppings falling to the ground. Because the oaks retain their leaves, I saw only dim suggestions of the porcupines. I wondered whether such an aggregation was a mating-season phenomenon or whether the cluster of oaks was a limited resource comparable to an apple orchard.

Only 10 minutes after my arrival, Mayday descended and set off noisily through the leaves. I gathered my gear and followed, reaching him at 4:15 a short distance south and at the same elevation. This time he had found his own oak, a vigorous tree growing in a clump next to a small clearing. He fed there for 20 minutes, then climbed a smaller oak tree in the same clump, his fifth tree for the night. After an hour of feeding and resting, he moved again. I followed sleepily, to a sapling-sized oak. It was now 5 a.m., and a first small grayness was stealing into the sky. Standing under the tree, I heard vigorous crunching above, then sounds of a porcupine descending, but the animal came down not from Mayday's tree but from a neighboring oak. An unmarked porcupine stopped about 6 feet above the ground, stared at me for a moment, then made a snorking sound like a child imitating a pig. It repeated the snork, then completed its descent and disappeared down the slope. Later in the morning, when I awakened from a nap under Mayday's tree, a red morning sun hung in the trees. Mayday was resting quietly. I started for home.

In mast crop years, the oaks draw porcupines through September and October, until the acorns fall to the ground and are rapidly cleared by deer, turkey, and chipmunks. There is concomitant feeding on the much smaller crop of beechnuts and on linden leaves as a nitrogen source. In 1983, 1986, and 1988, when the acorn crop failed, the oaks lay stripped by September, and the porcupines fell back on late-maturing apples and on linden.

The period of feeding on mast and linden is associated with continued weight gain, and adult porcupines reach their highest weights of the year. Juveniles maintain a positive weight balance up to the December snows by feeding on ground vegetation such as raspberry and green marsh grasses. By that time, adults have long been denned up or under hemlock cover, feeding on winter vegetation. They lose weight during the ordeal. Their survival depends on reserves accumulated through a summer and fall of hunting the rare leaves and rich fruit of the forest.

Notes

1. Correlation between leaf homogenate pH and feeding preference:
1984, $r=0.654$, df=11, $P<0.05$
1985, $r=0.737$, df=15, $P<0.01$
2. Apple seeds contain amygdalin, a cyanogenic gentiobioside (Conn 1979).
3. $t=3.25$, df=11, $P<0.01$.

7 The Winter Den

Winter's first thin coat of snow is a compulsive storyteller. Large and small mammals of the woods who have been skulking and stealing through the shadows are suddenly printed in the snow. Even after the animals are gone, their footprints remain to tell their stories. The density of animal populations revealed by such tracks is a surprise. Deer, fox, coyote, porcupine, wild turkey, squirrel, grouse, mice, and voles can all be seen reliably. The rare black bear, bobcat, skunk, or fisher is a lucky find. Hare and cottontail, once abundant, have all but disappeared. On such an afternoon of sparkling early-December snow I found a porcupine track crossing our marsh. It was my first encounter with this animal on its own terms. I spent 2 days following it in the snow from its origin to its destination, and I have been following others like it ever since.

A Trail in the Snow

A porcupine trail in the snow can be confused with the track of no other animal. The footprints are large and, like a miniature bear's, plantigrade. That is, the animal steps down on the sole of its foot, not just the toes. In contrast to the large footprint, the stride is short because the legs are short. The animal scuffs its way along, never jumping over a fallen log or a patch of water. There is a rhythm to the movement of the legs, as regular as a sewing machine stitch. When the track is fresh and the snow deep enough, one can see the overlay of a longer periodicity—the side-to-side sweep of the comblike tail.

My individual is easy to follow, because no other porcupines are abroad. The tracks of deer or wild turkey, by contrast, make a hopeless maze as one animal oversteps the trail of another of its kind. The porcupine crosses the marsh and heads gently uphill along an old sawmill road. Periodically it overtreads a deer path, the deer traveling in the opposite direction. The road grades down to Roarback Creek, where the creek flows in a deep, steeply banked gully. By following the path, the animal avoids the insecure footing of a straight descent. This is a well-used animal crossing. I can see a number of deer tracks following the same route. The creek's current in early December is relatively gentle, and I step across easily on projecting rocks. But the porcupine has crossed by a bridge I had not noticed—a fallen tree trunk angled across the stream. I can see its careful steps in the snow on top of the log. It then begins the serious effort of ascending the ridge on the opposite side. It climbs the ridge at an angle, lengthening its path but grading it for an easier climb.

After a while, the animal stops at a large beech tree to feed. But it abandons the tree after chewing off only a tiny test piece of bark near the ground, perhaps an inch square, and continues its journey. Its small body fits comfortably under a fallen tree around which I have to detour. It has urinated lightly in the snow. After passing under a tangle of wind-thrown branches, it begins a hard, frontal climb up the slope. The track ascends steadily, then burrows inside a large, hollow sugar maple trunk open at the bottom. The animal has moved in and out repeatedly, dragging out old leaves and other debris as though to clean out the hollow. It is looking for a den. But in the end, something about the site proves unsatisfactory, and the porcupine resumes its climb.

Higher up, it briefly explores another old sugar maple, a "General Sherman" tree, with two openings at the base. Winter winds would have whistled through the space. Near the tree, the porcupine intersects the trail of a solitary grouse, walking downhill and dragging its tail in the snow. An evergreen Christmas fern projects its green sawtooth edge above the white ground. The porcupine keeps climbing. It walks around a large-diameter beech tree that carries an old chew mark near the base, the work of another porcupine. It squeezes under a fallen tree, then crawls over another one concealed by snow. For a short time, it travels downhill along a lumber road, but soon it seems to recollect itself, leaves the road, and pushes uphill once more. It is entering porcupine country; more beech trees are displaying old chewing wounds on their trunks. The animal stops for a serious meal, chewing a sugar maple in two accessible places where the trunk forms sharply convex buttresses. Small fragments of outer bark litter the snow. After the meal, more climbing. I don't know how the animal feels, but I am overheating and have to open my winter jacket.

A deer track, more crawling under fallen trees, then drama in the snow. The porcupine track intersects the track of a red fox. The deer, squirrel, and grouse tracks gave no indication of porcupine interaction, but the fox is different. The two animals have clearly reached the same place at the same time and have exchanged greetings. Both tracks show milling and turning in the snow, although there are no loose quills or fur to indicate a physical attack. Then the tracks separate and resume their original directions of travel. It is easier for the porcupine to eat trees than it is for the fox to eat porcupines.

I follow the porcupine. More deer tracks, two sets of squirrel tracks, and then my goal—the lair of the porcupine. The animal is inside its chosen den tree, a strange old sugar maple like a giant bird's leg kicking skyward. The thick trunk carries a single eccentric branch at the very top. An opening at the bottom is too small for my head but wide enough for an incautious arm. I reach as far as I can but feel only space above. The porcupine is higher up inside and out of reach. Outside the opening is a spot of urine, pungent and porcupine-like, and two small clusters of scats melted in the snow.

The trail to the den tree has been a virgin trail; each step has been sharply recorded in fresh snow. The den tree is almost at the top of the ridge, and the trail climbs no higher. But a short, curious snow trail follows the isocline to the left. The footprints look blurred, and the ones I can read clearly are leading toward the tree, not away from it. I follow the trail for a few dozen feet before unraveling the mystery.

The tracks are fuzzy not because they are old but because they are superimposed on one another. One set leads away from the den tree; the other, on top, leads toward it. The porcupine has made a round trip and has traveled in its own footsteps. The object of the trip stands less than 100 feet away—a young rather twisted beech tree about 16 cm (6.5 inches) in diameter. High in the tree, about 10 square centimeters of bark have been chewed away. Above that, a branch has two large bare areas along the upper surface. The snow beneath the tree is scattered with the frizz of rejected outer bark. The animal has used only the inner layer—the phloem and cambium. The porcupine is gone, but the tree shows many feeding scars from years past, most high in the tree. Around the feeding tree, I count 19 other beech trees in a variety of sizes, and 14 show clear marks of porcupine feeding, mostly in the upper crown but often along the trunk. Most of the feeding scars are from past years. One medium-size tree 25 cm in diameter has been girdled sometime during the past 2 or 3 months; a 1-foot strip of trunk has been exposed all the way around the base. The tree is doomed. My first reaction is shock. I get a different perspective a minute later.

At the top of the ridge, I find a scene of destruction. A lumbering operation has reduced almost every sizable tree to a stump. Decaying slash lies tangled everywhere, with thick, young growth squeezing through to reclaim the forest. A super-porcupine has been at work here, one who does not eat the trees he cuts.

On my way back, I pause again at the den and look out across the ridge. The sun has set behind Vly Mountain but still glows red on Cave Mountain and the Jewett range, Beehive and Saddleback mountains. The landscape projects an incongruous sense of the Grand Canyon: great distances, somber red colors, and the dark shadow in the foreground creeping out across the valley. As often as I have looked out from porcupine dens in the mountains, I have found spectacle. But now the shadows make me hurry on.

To complete my picture of the porcupine's journey, I return the next day to the marsh and track the animal backward to its point of origin. The trail leads me across the stream to a small grove of hemlocks and then to the specific hemlock where the animal started. The evergreen branches of the hemlocks had given shelter during the snowstorm. From this lovely spot, the animal has moved constantly uphill, crossing a stream and a marshy area twice, moving about 0.75 mile in linear distance and 500 feet in elevation in one long trip, investigating den possibilities along the way and finally choosing one that is both comfortable and close to food.

I return to the den tree 36 days later, this time without a porcupine to guide me. I find all the landmarks along the way: the animal-crossing site across the Roarback, the General Sherman maple rejected as a den site, the sugar maple with the two basal feeding scars (now considerably darkened by exposure), then the feeding area of beeches with twisted tops, and finally the kicking-leg den tree with the small opening at the bottom. No tracks lead from the den tree—fresh snow has fallen, and a thin layer has infiltrated the opening. The porcupine has moved, and I am unable to follow.

I have spent a long time describing the den-searching porcupine because it revealed to me, in my very first encounter, many of the aspects of porcupine winter behavior in the Catskills. Every element of the story has proved to be significant: the starting point in the hemlocks, the long journey accomplished in a single day, the precise knowledge of local geography, and even the little drop of urine in the trail.

In wildlife research, we follow a subject with a will and personality of its own. We must always wonder whether the behavior observed is typical of the species, typical of the individual animal, or not typical of anything, as was a deer track I once found showing a fall on snow-covered ice or a raccoon I watched sniffing the tail of a skunk. To evaluate an observation, we must repeat it many times, test it for deviation from random process, and try to see it from the perspective of the organism involved. In the years since encountering the den-seeking porcupine, I have followed hundreds of similar tracks in the snow. Some of my questions have been answered, but a crop of new ones has grown up.

Figure 7.1. The trajectory of each footstep along a fresh porcupine trail is etched in the snow. (Photo by UR)

Den-Entering Time

Raccoons, beavers, and woodchucks use shelters year-round. My Catskills porcupines, on the other hand, are den dwellers only during the winter. In that sense, my den-seeking porcupine was out of step with the rest of its population, which had entered winter dens around the first week of November. Some clues to the factors that determine when a porcupine enters its den may be found in sporadic summer denning behavior. Almost all the time, porcupines in the summer months are found out in the open: in a tree when resting or feeding or on the ground when traveling. But on rare instances they may be found in dens (Figure 7.2).

On a Mother's Day in early May, my wife, Steph, and I went out to look at the woods and to locate two of our porcupines. The morning was splendid, with thick tree buds casting a dusky light and the first half-open leaves dancing in the sky. Only one factor spoiled the walk—newly emerged blackflies were biting hard on every inch of exposed skin. The female Squirrel was in a well-frequented resting tree, linden number 766. Instead of lying on a branch, she was inside a hole in the tree. My binoculars revealed her hunched up to hide all bitable skin. As I studied her, a blackfly bit my hand, and I watched

Figure 7.2. Early summer view of the opening of a porcupine den. While columbines and Canada violets bloom in the foreground, the resident has been absent for many weeks; it will not return till November. (Photo by UR)

the skin swell up immediately. Later on our walk, we discovered the large male Mayday in a similar position, in a hole in an old, leaning sugar maple. Fortunately for us and the porcupines, the blackfly season in the Catskills is short. But while the season lasts, the flies are fierce enough to drive at least some porcupines into dens.

However, Comtois and Berteaux (2005) could not demonstrate any connection between summertime den use and biting-insect severity in their Quebec porcupines. They speculate that their animals used dens both for fisher avoidance and for blackfly avoidance. As a result, the animals were found inside rock dens 31% of the time during the summer, whether blackflies were biting or not (Morin et al. 2005).

Another temporary denning stimulus is rain. The next time I found both animals inside tree dens was almost a month later, in early June. It had rained all day, often hard, and I had held off my tracking in hopes of clearing weather. The rain never stopped, and I went out in late afternoon. Again, I located both animals inside hollow trees. They were probably drier than I was. On another June day years later, a friend and I went out radiotracking during a light drizzle. We found Squirrel lying on a comfortable branch in a favorite resting tree, a large, hollow sugar maple. Suddenly the rain began pelting down. Squirrel got up, turned around, and climbed inside an opening high up the tree.

The North American porcupine is a widely dispersed, ecologically adaptable species. Its habitat encompasses not only the familiar woodlands of the north but also portions of the desert, such as the Mojave of California. It has been suggested that porcupines survive in the low-desert habitat by hiding from the daytime heat in dens, then venturing out at night to seek green vegetation (Betancourt et al. 1986). Even in the cooler Great Basin Desert of northwestern Nevada, porcupines often wait out hot summer days in thermally protected dens (R. Sweitzer, personal communication); that is the strategy of most desert-dwelling mammals. But the den can protect a porcupine against winter cold as well. Heat stress is not a problem in the Catskills; cold stress can be. A number of times I have found my animals driven into temporary dens by sudden cold spells. On the night of October 6, 1984, the temperature on the mountain crashed to an unseasonable 24 °F. Next morning, while patches of frost still whitened the ground, I found the females Moth and Cameo inside separate rock dens and Squirrel inside a tree den. A fourth animal, the juvenile Becky, was inside the hollow base of a sugar maple. Three other animals, whom I reached only in the afternoon when temperatures had climbed, were out in the open.

Prewinter denning responses to cold differ in three ways from winter den use. First, prewinter den use is episodic. Subfreezing temperatures drive most porcupines into dens, but they emerge again once the thermometer climbs. In the winter, animals typically remain in dens even when day temperatures rise above freezing. Second, prewinter dens are havens of opportunity, found near favorite roosting or feeding sites. With few exceptions, they lie outside the animals' winter territories and are never used in winter. Third, there are differences in den types. The great majority of prewinter dens in the Catskills are hollow trees; most true winter dens are rock crevices.

The stimulus for entering permanent dens appears to be a string of days when temperatures drop below freezing (Roze 1987). Data from six winters (Table 7.1) show animals entering permanent dens only after minimum daily temperatures have fallen below freezing for 7 days (range: 3–9 days). The average calendar date is October 29. Early cold spells in 1985 advanced permanent den occupancy to October 15. The particularly warm autumn of 1984 delayed permanent den entrance until November 10. The variation among years was greater than the variation among animals in a given year, suggesting a climatic response.

Table 7.1. Dates of permanent den entry (in calendar days) for den-using porcupines, 1982–1987

	1982	1983	1984	1985	1986	1987
	310	302	314	286	298	297
	302	295	315	286	315	284
		302	314	292	298	297
		303	314		298	304
			314		306	305
			314		297	304
						297
Average	306	301	314	298	302	298
Grand average	302 (October 29)					

ANOVA

Between years	df=5	SS=1611.9	MS=322.4	F_s=10.5	P=0.0001
Within years	df=22	SS=675.3	MS=30.7		

Den Types

At first sight, the winter dens chosen by porcupines seem to offer little protection from cold. Den trees typically have openings higher on the trunk, and the majority have an appreciable list, making for easy climbing. The opening, whether at the base or higher, is not closed. There is no insulation or nest material inside the den. On the contrary, porcupines may remove windblown leaves and debris. With only its tail presented to potential predators, the occupant can rest easy up inside the tree (Figure 7.3). But tree dens were used by my porcupines only 20% of the time.

A much more common den type, making up 70% of the winter dens used by my porcupines, is the rock outcrop. The Catskill Mountains are formed of sedimentary sandstone. In the valleys, a thick blanket of gravel and clay overlies the rock. At higher altitudes, the mountains' skin wears thin, exposing seams of sandstone alternating with conglomerate and shale. In this country of exposed rock and broken boulders, most porcupines spend their winters. Along the rock faces, frost and running water have bitten into the softer shale layers to create fissures. Those reaching 12 feet or deeper into the mountain are the den sites favored by porcupines. During winter, drifting snow covers the opening, leaving only a small entrance. With its back to the entrance, the porcupine sits in the deepest part of the recess (Figure 7.4).

The innermost region of the den is bare, though much of the rest of the floor is covered by the decomposing droppings of occupants from years past. Often the droppings are reduced to a fine sawdust and smoothed by years of porcupine traffic. In moist, temperate habitats such as the Catskills, much of

Figure 7.3. A porcupine in a hollow-tree den. Unlike this den, the entrances to most hollow-tree dens are higher up the trunk. (Photo by UR)

Figure 7.4. Squirrel's trail to Split Rock den. The jumbled footprints show she has traveled back and forth in her own tracks. (Photo by UR)

the physical breakdown of porcupine dung deposits is accomplished by a guild of mites, arthropods too small to be seen by the naked eye (Calder and Bleakney 1965). They also help to transmit certain porcupine parasites (Chapter 11). Where water is absent, as in desert sites, porcupine dung is cemented by the animals' urine into a concrete-like, durable conglomerate. Betancourt and coworkers (1986), who have studied porcupine dung accumulations in desert caves, report that the material could not be broken by hammer blows heavy enough to shatter the surrounding basalt.

The remaining 10% of the winter dens used by Catskills porcupines is split evenly between hollow logs and human outbuildings (barns, sheds, cabins). The buildings must be situated near the animals' food trees; they need not be unoccupied. One November night, at the height of the hunting season, I was expecting to find the porcupine Moth denned in a hollow linden tree in the middle of a hemlock forest. Instead, I tracked her radio signal to a cabin near our own woods. Two vans were parked outside; a dim light was burning. I formed a chilling hypothesis: hunters had killed Moth in the woods and taken her radiocollar inside. The radio's motion sensor had shown no sign of activity.

Early next morning, I drove to the cabin to confront the hunters. I pounded on the door. No answer. I pounded again. The door creaked open. To my surprise, the head that asked me what I wanted was only knee-high. A hunter had been sleeping on the floor and was addressing me from his sleeping bag. As I stood explaining, more and more bodies emerged. Finally, I was facing six men crowded in a little cabin about 16 feet square. They were not pleased to be awakened. They became indignant when I asked if they had shot my porcupine. To demonstrate that the collar was right inside the cabin, I turned on my radio receiver. But it was not; the signal was weak and came from a thousand yards away, where Moth lay sheltering under a barn after spending the night under the hunters' cabin. Perhaps the sounds of six hunters turning in their sleep had kept her awake. She never returned to the cabin den, for which I am thankful. I visited the cabin a week later, with a bottle of scotch and apologies. But the hunters had left.

Dens and Heat Conservation

If porcupines enter winter dens to escape from cold, their den choices seem curious. The dens contain no insulation, the entrances are open, and the animals don't huddle up for warmth. In addition, they emerge to feed at night, when outside temperatures are lowest. Without doubt, dens could be engineered to conserve more heat, but perhaps other considerations rule that out. Would an insulated den build up an unacceptable level of ectoparasites?

Would the insulation foul the quills? Would closing the entrance make sense if it had to be opened every night for feeding? Those questions have not been experimentally investigated.

But in important ways, the dens do protect porcupines against heat loss. Body heat may be lost in four principal ways: by evaporation, conduction, convectional flow, and radiation. Evaporation refers primarily to the unavoidable loss of water through the lungs. Conduction means the direct transfer of heat from one surface to another, as would happen if a porcupine stood on bare stone with its naked footpads. Porcupines minimize conductional loss by resting primarily in the sitting position, the body propped up by the tail, the front paws held against the chest, and the hind paws turned partly sideways so the footpads are not in direct contact with the ground. In convection, a fluid such as air or water transfers heat. The principle is familiar to us as the wind chill factor—we lose heat more rapidly on a windy day. The air inside a porcupine den is still. Perhaps that is why my den-seeking porcupine rejected the General Sherman tree with its two openings; it would have been drafty. Convectional heat loss would be lowest in rock dens, which have deep and often curving entranceways (Figure 7.5).

Heat loss through infrared radiation is analogous to the flux of light or ultraviolet energy. The amount of heat lost is proportional to the surface temperature of the radiant body and to the difference in temperature between the radiant body and the heat sink.[1] The excellent insulating properties of porcupine fur (Chapter 3) keep the animal's exterior temperature low, close to that of surrounding rocks and vegetation. Hence, radiational heat loss to the surroundings should be minimal. Nevertheless, the open night sky can have a far lower temperature and can therefore be an effective heat sink. Shielding from the open-sky heat sink should contribute significantly to heat conservation. Conversely, a bright sun on a cold winter day can deliver radiant heat to an animal sitting in the open.

In February 1988, I installed temperature-recording devices for 1-week intervals in a ground den and a hollow-log den. One recorder was placed inside the den, the other outside, out of the sun and one foot above the snow line. For the rock den (occupied sequentially by at least two animals during the winter), a high of –1 °C was recorded February 1, when the outside temperature reached 4 °C during a midwinter thaw. The den low of –6 °C was recorded February 7, when outside temperature fell to –19 °C (–2 °F) (Figure 7.5a). At the time, the inside-outside differential measured 13 °C (22 °F). For the hollow log (occupied by the male Jasper), the high of –3 °C was recorded February 20, when outside temperature reached –2 °C; the low of –11 °C followed late the following day, when outside temperature fell to –17 °C (Figure 7.5b). The maximum inside-outside differential was 9.4 °C (17 °F), observed February 21.

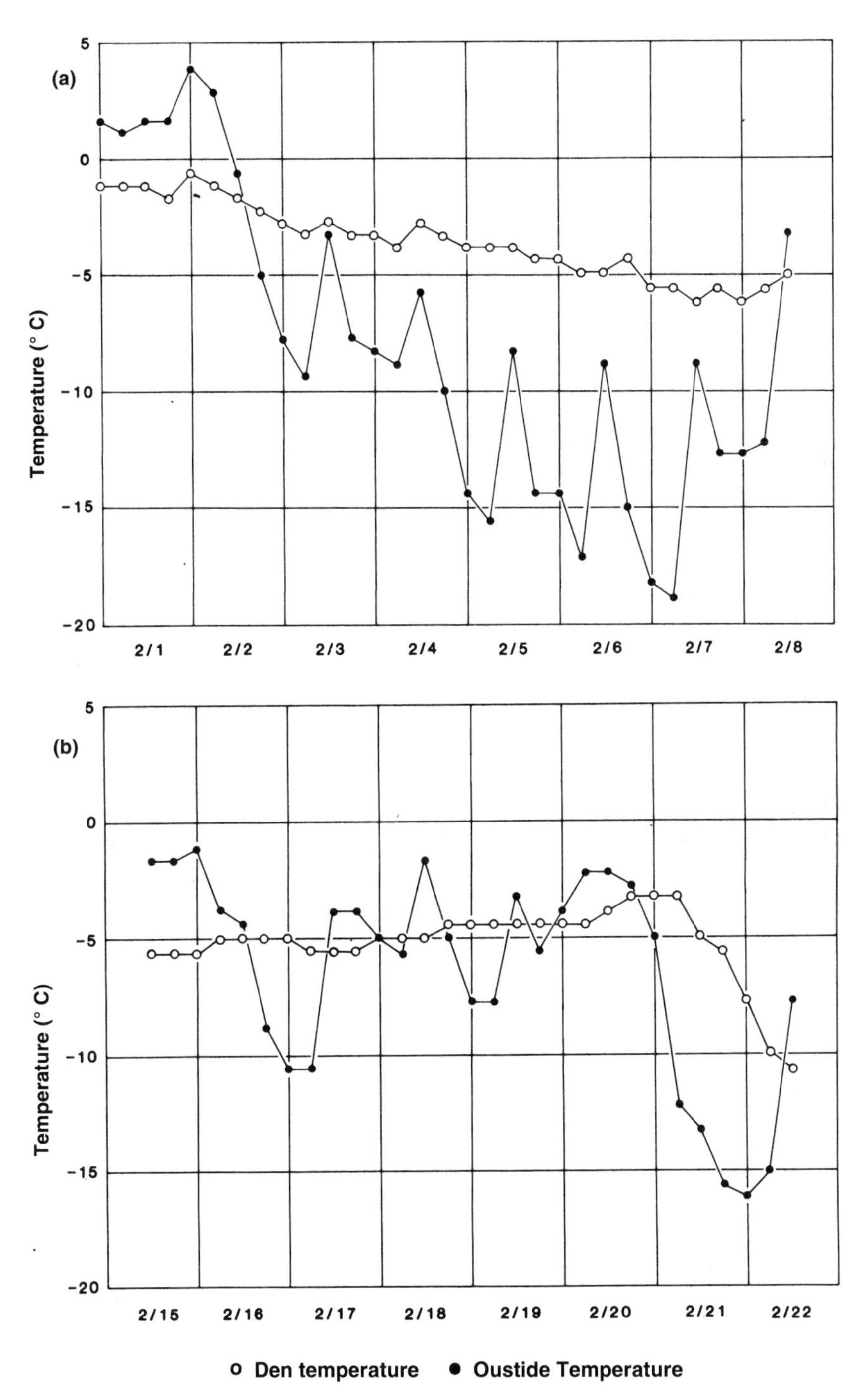

Figure 7.5. Internal and external temperatures in two den types: (a) rock den (b) hollow-log den. The differential between inside and outside is higher in the rock den, indicating better thermal protection.

According to Fournier and Thomas (1999), the critical temperature for porcupines is –2 °C (28 °F). When external temperature falls below that, the animals increase their metabolic rates above the basal level to maintain core temperatures of 37.5 °C. Based on the critical temperature given by Fournier and Thomas, neither den type offers full protection because den temperatures fell to –6 °C in the rock den and –11 °C in the hollow log. The rock den is superior, offering full protection for 1.5 days out of 8.

Both types of dens are essentially temperature-averaging devices, but the thermal mass of the rock den is much larger. In any case, both den types protect against convectional heat loss and radiational heat loss to the open sky.

But a simple temperature differential does not adequately explain an animal's heat conservation needs because it does not account for convection, conduction, or radiant heat exchange. To try to account for all of these factors, Mabille, Berteaux, and colleagues (in press), monitored winter heat loss by means of two artificial, internally heated porcupines.

The artificial porcupines contained copper cores (for efficient conduction) maintained at a constant 37.5 °C (a porcupine's winter core temperature) by means of external batteries whose drain was recorded electronically. The cores were covered by the pelts of two porcupines that had died naturally in the preceding year. The power required to maintain the core temperature then reflected the thermal drain of the artificial porcupines' environment.

In addition, the authors telemetered 16 porcupines to monitor patterns of den use over two Quebec winters. The following conclusions emerged:

- Porcupines saved an average of 16% metabolic energy by occupying their dens instead of open terrain. The savings increased at very low operating temperatures[2] (essentially air temperature plus wind chill). Thus, when operating temperatures dropped below –20 °C, the metabolic savings of den occupation rose to 25%.
- Over the course of two winters (December to April), the mean time spent outside the den was 6.7 h/day. At operating temperatures below –20 °C, time spent outside the den dropped to 5 h/day. The outside time rose as operating temperatures rose. At temperatures between 0 and –5 °C (roughly the porcupine critical temperature), males spent 50% of their time outside the den. Although low operating temperatures forced the animals to spend less time out in the open, the cold also forced them to spend more time feeding, presumably because of increased energy demand.

And how did the artificial porcupines of Mabille and colleagues perform in their coniferous closed habitat? At both high (>–5 °C) and low (–20 °C) operating temperatures, a coniferous habitat provided the same energy savings as a den. At intermediate operating temperatures, the artificial porcupines paid up to a 20% penalty for outdoor life, that is, they lost more heat than they would have lost inside a den. This helps explain a curious observation I made in my third winter of radiotelemetry—some porcupines spend little or no time in winter dens. For the first two winters of observation, all my animals occupied dens. In the third winter, I added the crippled male Imp to my study. He spent the winter in a hemlock-dominated area, entering his first den very late, on December 16. The den was the crawl space under a barn. The next week, Imp had moved to a neighbor's shed nearby. But he stayed under the shed for only 3 weeks and then left dens permanently. He spent the rest of the winter in the open, sheltering in or under hemlock trees.

The first time I found Imp in that situation was on a bitter night in January. The temperature had fallen to –28 °C (–18 °F), and a dry snow lay in drifts. He was under a hemlock, lying on the ground inside a crown of twisted apple branches. He had spent considerable time there. I saw several bedding hollows, such as deer make in the snow, and two large accumulations of droppings. A feeding trail moved off to the right. Imp looked moribund. He did not stir or erect his quills when I approached. A fine coating of snow covered his face, close to the ground. His condition looked so grave and the night was so bitter that I gave him only a 50-50 chance of surviving the night. When I returned to check on him the next day, I was so uncertain of what I would find that I carried a heavy plastic bag as a body bag.

I reached the large hemlock with the apple thicket at its base. Imp was not inside. I looked up into the tree, dark in the dusk. There, close against the trunk and about 25 feet up, was the black shape of my animal. I thought he looked triumphant. For the remainder of the winter, Imp was always either up in a hemlock tree or in some thicket at its base. His feeding trail in the snow showed the drag mark of his paralyzed foot. He froze to death in March during a severe cold wave, when temperatures fell below those of January.

Because Imp had been crippled, I ascribed his behavior to his injury. And because he froze to death, I assumed that wintering in the open was nonadaptive. But in subsequent winters, two normal animals, the adult males Homer and Killer, went through the same behavior without visible ill effect. For 2 weeks at the beginning of the winter, Homer occupied rock dens. He spent all the rest of the winter in or under hemlock trees. Killer spent two entire winters in hemlock cover. One difference between Imp and the other males is that the latter wintered in hemlocks that formed a thick, unbroken forest. Imp's hemlocks grew in scattered groves and formed a less effective weather shield.

Evergreen cover offers thermal advantages in the winter. Hemlock needles are layered so thickly in the canopy that they shield a porcupine from the heat sink of the open sky. The trunks and foliage of the hemlocks also act as secondary sources of radiant energy, reradiating at night the energy they have absorbed during the day as well as the long-wave radiation emitted by the snow and surrounding vegetation (Clarke and Brander 1973). Naturalists across the country report that porcupines may spend winters in trees, away from formal dens. In every report that lists the tree species, they have been evergreens. Brander (1973) reports that a Michigan porcupine spent 26 consecutive days sleeping and feeding in the crown of a hemlock. Smith (1982) in northeastern Oregon found his animals sheltering from snow and wind in large, mistletoe-infested Douglas firs. The growths created witches' brooms, whose dense architecture was particularly effective in sheltering the animals. Curtis and Kozicky (1944) list the most common "station trees" in Maine as hemlock and white spruce (Figure 7.6). In the 1930s, Walter Taylor (1935) described the porcupines of Idaho as spending much time in the winter in yellow pine, Douglas fir, and white fir.

Animals other than porcupines may depend on the hemlocks' green winter blanket. White-tailed deer converge on local hemlock stands during the winter and leave a maze of trails and sleeping depressions known as a deer yard.

Figure 7.6. Pins and needles; porcupine in a white spruce. This tree species forms an important winter food in the north. (Photo by UR)

Their choice is commonly ascribed to the reduced snow cover under hemlocks. Perhaps the hemlocks' sky-sheltering properties are of equal importance.

Spending the winter under hemlocks does not preclude a porcupine's use of a winter den. Moth used a series of hollow trees in a hemlock grove as dens. But she emerged from her dens earlier in the spring than other animals and spent the interval in hemlock shelter. At least three other porcupines awaited full spring in hemlock groves after abandoning their rock dens for the winter.

Den Fidelity

Because porcupines do not dig or excavate their own dens but depend on ready-made shelters, good dens are often in limited supply. That may be why porcupines return year after year to the same dens. Squirrel, my animal under longest observation, returned to the same palatial rock den for 20 winters. Every other animal that I observed for more than one winter returned to at least one den used in the previous winter. Sometimes there are breaks in the pattern of occupancy. Rebecca used den A in 1982–1983, but not in 1983–1984; she returned to it in 1984–1985 and in the three subsequent winters. Animals returned to both rock and tree dens; outbuildings had less of a magnetic effect, perhaps because they were less intensively used to begin with. Other researchers have suspected but never proved that porcupines return to dens. In Wisconsin, Krefting and colleagues recaptured a female in December 1952 that had been captured and marked 2 years earlier only 100 meters away (1962). But he also found other animals far from their sites of initial capture.

Den Change

Krefting's discrepancy is explained by porcupines' regular moves to new dens throughout the winter. I found my animals in new quarters, on average, every 23 days. Thus, each animal was associated not with one den but with a suite of dens. The reasons for their frequent moves remain obscure.

One factor that regularly induced movement was a disturbance at the den, such as trapping the animal. Of 19 cases when animals were trapped and promptly released at their dens, den change followed in 12 cases. The changes were usually not permanent.

Disturbance does not account for all the observed moves. Adult males may have greater wanderlust than juveniles or females. Of three adult males I followed for at least one winter, two did not use dens. The third adult male,

Finder, used normal dens but changed them every 9 days. That rapid rate of change differed not only from the patterns of females but also from Finder's own pattern as a juvenile, when he changed dens at the same rate as females and other juvenile males. Whether the pattern is true of other adult males requires further investigation.

Also requiring further study is the role of food in den change. Do animals change dens because they exhaust the food supply in the vicinity? On its face, that seems unlikely because porcupines often return to vacated dens later in the winter and resume feeding on the same trees they were gnawing before. What cannot be ruled out is that attacked trees throw up a chemical defense against the tree predator, just as they do in response to summer defoliators. The defense may be only temporarily effective and allow the animal to return weeks or months later. The question has not been investigated.

Den Sharing

For most of the winter, porcupines occupy their dens alone, sitting in their dimly lit recesses by day and emerging to feed at night. Activity indicators on their radiocollars suggest that they move little or not at all while in their dens. That dozing, banked-down existence may be broken by the arrival of a second porcupine.

One cold December morning, Squirrel's radio signal led me to her favorite winter den, Split Rock den (see Figure 7.4). Instead of the customary single feeding trail in the snow, two trails emerged from the den entrance. One led downhill to a near clump of freshly chewed sugar maples; the other climbed steeply uphill. As I stood examining the feeding sign, high-pitched squawks from within the den broke the winter stillness. They were the cries of an annoyed porcupine, and they did not let up until the beam of my flashlight pierced the interior gloom of the den. In the back of the long passageway stood two porcupines. Squirrel was in the back, apparently barring the passageway. The other animal carried no radiocollar. When I returned a week later, Split Rock den was silent, with Squirrel alone in possession of her spacious quarters.

On average, my porcupines shared winter dens only 12% of the time. There are clear indications that the experience is unsettling to both animals. Porcupines normally urine-mark the snow outside their dens, leaving a series of drips or small dribbles in the middle of the feeding trail. When two animals are in residence, both mark their trails, and the dribbles may grow into ostentatious streaks. Den-sharing seldom lasted longer than a week. In 12 of 17 instances of sharing, both animals abandoned the den, reacting to the sharing as they would to a serious disturbance. Other investigators report similar results. Jacob Shapiro (1949) in the Adirondacks never found more than one animal in a den. Wendell Dodge and Victor Barnes (1975) in western Washington report simultaneous occupation of a den by two porcupines only four times, and then only during breeding season. Their study site had a rugged topography with an abundance of den sites. Robert Brander (1973) in Michigan reports dual den occupancy in only 3 out of 46 cases. However, Dodge's (1967) porcupines in western Massachusetts shared dens so often that he describes the animals as "gregarious" during the winter. At the same time, he cites evidence of strife—the animals could be heard squawking inside their dens. The most probable explanation for those anomalous results is that Dodge's study site around Quabbin Reservoir contained few acceptable den sites. The rock outcrops in a rolling countryside contain few cavities, and the shortage may have forced porcupines to share dens.

Griesemer and colleagues (1996) investigated porcupine den use around the same reservoir 30 years later. They compared den use in two areas, one with twice the density of rock and tree dens as the other. In the den-enriched area, den-using porcupines shared 23% of the time. In the den-poor area, den users shared 71% of the time.

In about half the cases in my study, only one of the two animals could be identified by its radio signal or by capture. But in the 6 cases out of 11 when both animals could be identified, they formed a male-female (MF) pair. The odds of that happening by chance alone are less than 1%.[3] Griesemer and colleagues (1996), however, found no such sexual preference. Their den sharers made up MM (male-male), FF (female-female), and MF pairs, as well as FFF, FFFF, and FFM, FFFM, and FFFFM multiples. Interestingly, neither low temperatures nor deep snow promoted huddling.

The reluctance of porcupines to share dens may act as a population-dispersing mechanism in late fall, when animals first search for dens. As a young female wanders in search of an unoccupied tree or rock crevice, she may leave her birth site far behind. She will spend her first winter alone.

Den Abandonment

The days lengthen. Icicles grow over the entrances of rock dens and drip in the warming sun. The snow takes on a coarse, granular texture, and openings melt out around tree trunks. A red-eyed vireo begins to call. The time is coming for porcupines to abandon their dens. In the weeks of late winter or early spring, they can be found on windless days sunning themselves in trees outside their dens. Then, one day, they are a great distance away, feeding on new food sources and embarked on a new cycle of seasons.

The stimulus for den entrance was a string of days with minimum temperatures falling below freezing. The stimulus for den abandonment appears to be a combination of at least two factors: daily temperature minima must again climb above freezing, and the snow cover must melt off sufficiently to allow easy travel. If either requirement is not satisfied, the animals stay in their dens.

In 1983, temperatures had climbed above freezing by the first week of April, and the snow cover was shrinking daily. Then heavy snow fell for 2 days in mid-April. The snow pack reached its maximum thickness for the winter and did not disappear again until the last day of April, when temperature minima were well above freezing. All porcupines waited in their dens until the snow was gone. All were found out of their dens at the same time, on April 30 (Roze 1987). In 1985, the opposite situation developed. In a winter with only 35% of normal snowfall, the ground was bare by the third week of March. But porcupines held their places in dens until the first week of April, when minimum daily temperatures had climbed above freezing. The 1983 season saw the latest date of dispersal from dens; 1985 saw the earliest. But in both years, the same conditions had been achieved: a temperate, snow-free environment. Even as the porcupines left their dark, often cramped quarters, each animal remembered the spot well because it would be coming back in another winter.

Notes

1. Radiant heat loss, in kilocalories per square meter per hour, is given by the equation $Q_r = \varepsilon\sigma T_s^4$, where ε is the emissivity, ranging from 0 to 1; σ is the Stefan-Boltzmann constant; and T_s is the surface temperature in degrees kelvin (Robbins 1983).

2. $T_o = -3.638 + 1.0504T_a + 1.5757(\text{wind})^{1/2} + 0.0067(\text{net radiation})$, where T_o = operating temperature, T_a = air temperature (Mabille, Berteaux, et al. in press).

3. $(2 \times 0.34 \times 0.66)^6 = 0.008$.

8 Winter Foraging

On a January afternoon, I walk just below a ridge of Vly Mountain, as wind snatches away the sound of my snowshoes on the snow. A sudden rasping sound makes me pause. Not 20 feet away, a porcupine is debarking a small beech tree. It is about 12 feet off the ground, feeding on a 2-inch-diameter trunk, working from the top down. A 5-foot expanse of bare wood shines above it. The animal is removing the thin bark all the way around the branch; no portion is being rejected. It keeps changing the angle of attack as it chews rhythmically, then seems to lick the chewed area, or perhaps just swallows a mouthful of bark. As I approach slowly, it continues working. It may not have heard me or may not be used to humans walking along the ridgetop. Now I notice that the tree it is working on is stunted and misshapen, as are the neighboring beech trees. This is the work of porcupines in winters past. Only when I stand directly below the porcupine and a human body length away does it take notice. It stops feeding, moves to the opposite side of the trunk, and erects its quills.

Leaving the porcupine to compose itself, I follow its snow-drifted feeding trail to a rock den 40 feet downhill. A flat, deep entranceway disappears inside the mountain. The view from the cave entrance is framed by large trees on both sides. Saddleback Mountain rises in the background. Below it lies the dark hemlock forest, bordered by white fields and a twisting country road. Through windblown snow crystals, the mountain has been joined to the earth and sky. Perhaps from a porcupine's vantage point, a wandering human in this landscape is too minor an element to react to.

Feeding Diversity

My hilltop porcupine was atypical in only one respect—it was feeding during the daytime. But, like others in the area, it was a chewer of tree bark, a maker of snow trails, and an invader of available shelter. The winter food of porcupines is easily identified; the patches of missing bark in the tops of trees can be spotted from hundreds of feet away. In addition, they may use twigs, buds, and the needles of evergreens (Figure 8.1). Because of the visibility and economic importance of their feeding mode, porcupine winter food choices have been amply documented. The specific makeup of the winter diet seems to vary from community to community. For example, Tenneson and Oring (1985) compared porcupine feeding patterns at two sites 3 kilometers apart in Itasca State Park, Minnesota (Table 8.1). The French Creek porcupines used white pine, elm, and linden for 90% of their diet. For their Green Lake neighbors, the same species accounted for only 23% of the diet. It is significant that the three food species made up 24.8% of the forest at French Creek but only 10.4% at Green Lake. With populations only 3 kilometers apart showing such diversity of feeding behavior, it is not surprising that the porcupines of Idaho, New Brunswick, and Arizona should show even greater differences in winter food preference (Table 8.2).

Figure 8.1. Porcupine feeding sign in hemlock. Hemlock bark is not used because of its high tannin content; instead, animals nip off terminal branches (preferably those exposed to the sun) and consume needles and small twigs. (Photo by UR)

Selectivity

Glancing at Table 8.2, we might conclude that porcupines are not selective about winter food choice. That would be wrong. In the Catskills, sugar maple is a preferred winter food, while red maple is almost never touched. Likewise, white ash, whose young spring leaves are highly preferred, are never attacked for their bark. Hop hornbeam, a common species at lower elevations, is never sampled in the summer and used only sparingly during winter. Clearly, porcupines are both selective and opportunistic in their winter food choice.

Table 8.1. Major winter food choices of porcupines at two sites in Minnesota

	French Creek		Green Lake	
Tree	in forest (%)	in diet (%)	in forest (%)	in diet (%)
White pine	6.7	50	1.1	19
American elm	5.8	20	0.8	0
American linden	12.3	20	8.5	4
Quaking aspen	0	0	11.4	14
Red oak	0	0	3.0	41

Source: Data from Tenneson and Oring (1985)

Table 8.2. Major porcupine winter foods in North America

Food species	Location	Source
Spruces, white cedar	New Brunswick	Speer and Dillworth 1978
White spruce	Northern Quebec	Payette 1987
White cedar, hemlock, beech	Maine	Curtis 1944
Hemlock, beech, sugar maple	Massachusetts	Dodge 1967
Hemlock, white pine, red oak	Massachusetts	Griesemer et al. 1998
Hemlock, red spruce	New York Adirondacks	Shapiro 1949
Beech, sugar maple	New York Catskills	Roze 1984
Jack Pine	Ontario	MacDonald 1952
Hemlock, sugar maple	Michigan	Brander 1973
Sugar maple, yellow birch	Wisconsin	Krefting et al. 1962
Yellow pine, Douglas fir, piñon	Arizona	Taylor 1935
Ocotillo	Arizona desert	Reynolds 1957
Gambel oak	Utah	Stricklan et al. 1995
Ponderosa pine	Idaho	Curtis and Wilson 1953
Douglas fir, limber pine	Alberta	Harder 1979
Yellow pine	Oregon	Gabrielson 1928
Spruces	Alaska	Murie 1926

Nutritional Ground Rules

Some general principles can help rationalize the complex and contradictory picture of winter feeding behavior. As Daniel Janzen of the University of Pennsylvania has paraphrased the common proverb, "One beast's drink is another beast's poison." A given plant species is not universally acceptable to all animals. Red maple twigs, rejected by porcupines, are important winter browse for white-tailed deer. Apple bark, seldom touched by porcupines, is eagerly sought by field voles and cottontails; apple twigs are browsed by deer. Likewise, white ash twigs attract deer, who cannot reach them in the trees but travel long distances to feed on windthrows and logging slash. (The greater range of species options available to deer is counterbalanced by their inability to climb trees and chew bark.)

Every plant–herbivore system is a mutually interacting system. Plants respond to attack by erecting antiherbivore defenses. Examples of evolutionary plant defenses are the physical defenses such as thorns, silica crystals, and leaf hairs. Chemical defenses may be either long-term (evolutionary) or short-term (ecological) in nature. They include compounds such as tannins, terpenes, alkaloids, cyanogens, salicylates, and a host of others. A herbivore can use the stored resources of a plant only if it can overcome the plant's physical and chemical defenses. Animals differ little in their basic metabolic requirements. They differ enormously, in both qualitative and quantitative ways, in their ability to overcome plant defenses.

In mammals, the primary organ of detoxification is the liver. Here toxic molecules are broken down or conjugated with carriers such as glucuronic acid to make them suitable for excretion. In herbivorous mammals, gut microorganisms play an equally important detoxification role. One spectacular example comes from the work of Allison (Allison et al. 1987), a U.S. Department of Agriculture (USDA) microbiologist in Ames, Iowa, and Jones and Megarrity (1986) in Queensland, Australia. They noted that goats in Hawaii could safely feed on *Leucaena leucocephala,* a small leguminous tree that produces the toxic amino acid mimosine. Cattle in Florida and Australia were poisoned by the same tree. Allison and his coworkers showed that a unique bacterium in the goat rumen degrades a toxic mimosine metabolite known as 3,4-dihydroxypyridine (DHP), protecting the goats. When Australian steers were inoculated with rumen fluid from the Hawaiian goats, they acquired the same DHP resistance as the goats.

The gut microorganisms of ruminants such as goats and deer inhabit the rumen, situated at the head of the small intestine. In porcupines, the gut microorganisms are concentrated in the caecum, at the end of the small intestine.

That positional difference as well as qualitative differences in gut microflora may account for differences in food tolerances between porcupines and ruminants. In addition, indirect evidence suggests that the gut microflora may vary from one porcupine to another.

Detoxification is only part of the task of the gut microflora. Its other function is digestion. Plants lock up much of their carbon stores as structural carbohydrates, in cellulose, hemicellulose, and pectin, which are inaccessible to the digestive enzymes produced by the mammals themselves.[1] Hence, all herbivores depend on the digestive enzymes of their gut microflora. The cellulose-digesting ability of porcupine caecal bacteria was demonstrated in 1949 by Balows and Jennison.

The porcupine's diet must supply two necessities: energy to drive the life processes (derived in significant measure from cellulose digestion) and nutrients to replace worn-out body components. Most important of the nutrients are nitrogen (needed to build proteins and nucleic acids) and certain minerals. In terms of nitrogen, the porcupine's diet is distinctly inadequate. Better-quality winter foods contain nitrogen in the range of 1% by dry weight (6.2% crude protein). Most tree bark falls below that level, with 0.5% a typical value. Animals cannot maintain constant body weights at that level of nitrogen intake. As a consequence, all porcupines I have examined lost weight during the winter. They survived only because they had built up reserves during the summer and fall. Similar data are observed by Sweitzer and Berger (1993) and Berteaux et al. (2005).

If nitrogen and metabolizable carbon are low in winter foods, why don't the animals simply eat more? There is no lack of bark in the forest. The problem is processing time. The bacterial digestive processes of the caecum are relatively slow; it takes approximately 2 days for food to pass through the digestive tract. A faster passage would not give caecal bacteria sufficient time to do their work.[2]

Feeding Technique

Porcupines typically feed at night, although warmer weather brings them out in the day as well. With the settling of dusk, they snowplow their way toward a feeding tree, usually less than 300 feet away. Sometimes the animal feeds at the base of the tree, but more commonly it climbs into the crown and eats the bark of upper branches. With thick-barked trees such as adult sugar maple, it feeds in two phases. A first pass of the incisors shaves off the dead outermost cork layers—they fly in all directions and fall on the snow below (Figure 8.2). On the second pass, the porcupine harvests the inner bark, shaving it down to the cambium, grinding it with the cheek teeth, and swallowing it. Usually,

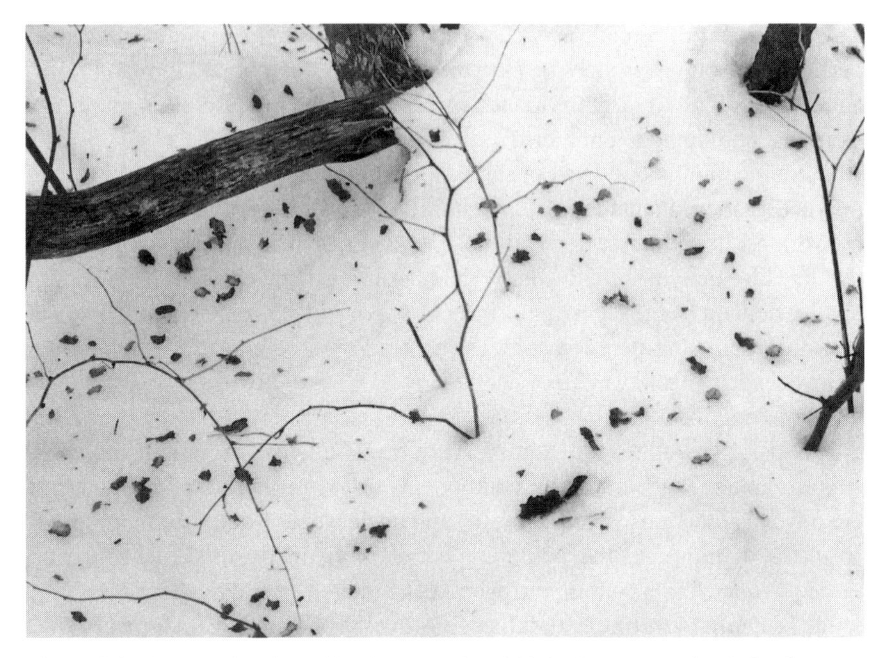

Figure 8.2. Outer bark, rejected by the porcupine, falls in the snow. Animals feeding on thin-barked branches consume the entire bark. (Photo by UR)

bark on the outer branches is thin enough to be harvested in its entirety, in a single pass.

The pattern of tooth scrapes left on the trunk by an adult porcupine displays the animal's craftsmanship. The bark is removed only down to the cambium; the remaining surface is plywood-smooth, more finely finished than the work of beaver. The porcupine removes the bark in small triangular patches, each patch composed of five or six scrapes converging on an apex, like sticks in a tepee. The apex represents the position of the upper incisors, held fixed against the bark. The lower incisors scrape, moving over a fresh path as the lower jaw swivels in a narrow arc. Examined carefully, the anchor points of the upper incisors can be seen as spaced indentations around the edge of the open scrape. The final pattern on the trunk conveys rhythm, economy, and elegance. The animal does not scrape the same surface twice, does not leave residual strips of bark, and does not vary the depth of its scrape.

As I walked one day through persisting March snow, I came across a pair of chewed yellow birch that violated all those norms. The feeding job was strikingly messy. The animal had wasted effort by going over some areas twice; had elsewhere left scraps of inner bark on the trunk; had scraped at varying depths; and showed a chaotic, unrhythmic attack. A faint feeding trail led to a gray boulder on the edge of a clearing. I peered into a shallow recess underneath. At the far end, the darkly furred form of a small porcupine lay pressed against the rock. Capture was easy—the juvenile male was too small to put up a fight. It weighed 1.9 kg (just over 4 pounds) and carried one of my tiny numbered tags in its right ear. I had tagged it the previous summer, when it was only a few weeks old. It was now spending its first winter on its own. Like many a human baby, it remained a messy eater (Figures 8.3 and 8.4).

Feeding Specializations of Individuals

The porcupine's feeding trail in the snow is a notebook of information. When I began to examine a series of such feeding trails, I was struck by something

Figure 8.3. Juvenile's feeding sign—a messy job. (Photo by UR)

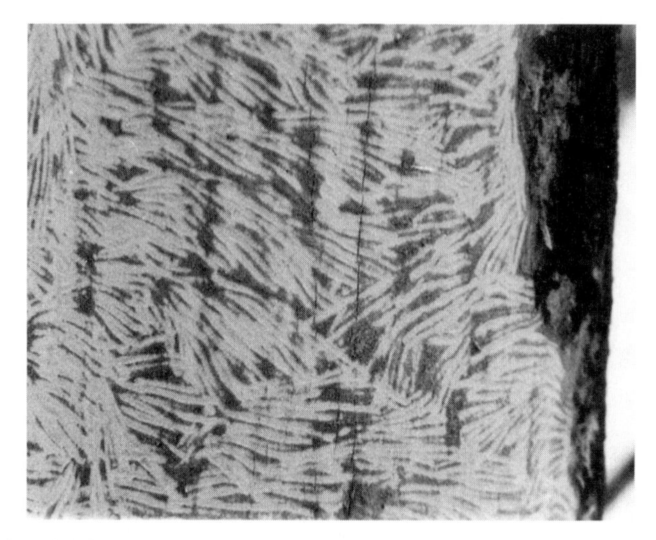

Figure 8.4. Adult feeding sign—even, rhythmic tooth scrapes. (Photo by UR)

curious. Each animal was feeding on only one (sometimes two) species of trees, encountering and rejecting trees along its feeding trail that other porcupines might find entirely acceptable, might in fact be using as their species of choice. Figure 8.5a illustrates the distribution of trees along an actual feeding trail. The circles represent 17 trees attacked by the porcupine, some with only a light test bite, others heavily chewed. One small tree, 6.6 cm in diameter, had been stripped bare from top to bottom. All 17 trees were of a single species, beech. I counted all trees inside the trail, plus all trees within 3 meters (about 10 feet) around it, and assumed that they were available as feeding choices to the porcupine in question. The area encompassed 158 trees, with 87 beech, 37 sugar maple, 28 hop hornbeam, and 6 white ash. Of those four species, beech and sugar maple are highly preferred winter foods for the porcupine population as a whole, hop hornbeam is used rarely, and white ash is never consumed. The odds of attacking 17 beech trees by chance alone are less than 5/1000 (0.005), much lower than biologists like to accept as being due to random process.[3] It is more reasonable to assume, therefore, that the porcupine is actively selecting beech to feed on.

Now consider another porcupine, observed a year later and about 1 mile away. This animal picked its way downhill, traveling in the snow shadows of two large boulders and a rock outcrop, then fed on 7 sugar maples strung out on a line (Figure 8.5b). The 28 trees within 3 meters of the feeding trail included 16 sugar maples and 2 each of red maple, striped maple, mountain maple, hop hornbeam, beech, and white ash. Porcupines regularly feed on

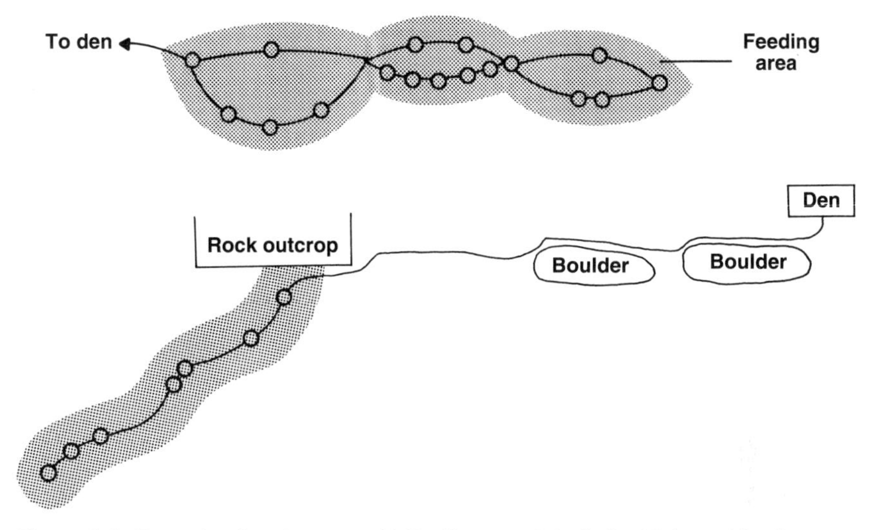

Figure 8.5. Porcupine foraging areas. (a) Feeding area (stippled) with looped feeding trail; all trees selected are beech (b) Feeding trail (stippled); all trees selected are sugar maples.

sugar maple, striped maple, mountain maple, and beech. Hop hornbeam and red maple are rarely touched, and white ash is rejected. The probability of randomly selecting 7 sugar maples in this feeding area is greater than 0.5, not low enough to rule out random selection.[4] Nevertheless, the second porcupine selected a diet that did not overlap the diet of the first porcupine.

The two porcupines discussed here (and all others I have examined) selected feeding areas in which their preferred tree species was the most abundant species. The first porcupine, feeding on beech, encountered beech 55% of the time while feeding. The second porcupine, feeding on sugar maple, encountered sugar maple 57% of the time. That could mean that a porcupine's feeding choice is dictated by surrounding forest composition, or it could mean that porcupines select areas enriched in the preferred tree species. To explore the question, I selected control areas for each feeding area and statistically sampled their species composition. The control areas were located 30 meters from the feeding areas and at the same elevation, and they measured 30×30 m; 100 trees were sampled in each. Sample comparisons of forest composition in feeding and control areas are given in Table 8.3. Because the control areas were larger and more evenly distributed, they reflected more accurately the local forest composition. The comparisons suggest that porcupines are selecting patches of forest that are richer in the desired food species. The differences are statistically significant and are large enough to reflect active choice, not random fluctuation in forest composition. Why should the porcupine make

Table 8.3. Tree distributions in feeding and control areas of two porcupines

Species	Number in feeding area	Number in control area
	Porcupine 1	
Beech[a]	87 (55%)	31 (31%)
Other	79	69
$\chi^2 = 14.29$, df $= 1$, $P < 0.001$		
	Porcupine 2	
Sugar maple[a]	16 (57%)	33 (33%)
Other	12	67
$\chi^2 = 5.40$, df $= 1$, $P < 0.025$		

[a] Primary feeding choice

that choice? Perhaps for the same reason it travels in its own footsteps and does not venture far from the den to feed: its body is poorly designed for travel in deep snow. By selecting a winter foraging area where the desired species is dominant, it can reach its food in fewer steps.

But why should an animal narrow its winter diet to only one or two tree species when the total population feeds on at least ten species? Two answers may be considered. If two porcupines feed on entirely different tree species, as was true of the porcupines in the preceding example, then the two animals should coexist without competition, and the forest can support a larger total population. But my animals seldom shared dens and feeding areas. The only exceptions were male-female pairs who had previously shared summer territories. Whether such cohabiting animals had the same or different food preferences remains to be determined.

A second possibility might be titled the specific microflora hypothesis. The gut microflora must, in part, detoxify plant defense compounds in the porcupine's food. If different tree species carry different defense compounds, it might be to a porcupine's advantage to specialize on species that its own microflora are best able to detoxify.[5] Failure to specialize might lead to poisoning by plant defense compounds or might lower digestive efficiency. Because the margin of survival during winter may be thin, even a small decrease in efficiency might be fatal. Although this hypothesis has not been tested directly, it can be tested indirectly by following diet changes as animals move from den to den. If an animal's feeding choices are innately fixed, they should not change at a new den. If, on the other hand, feeding specializations simply reflect local tree abundances, then animals who change dens most often should show the greatest diversity of feeding behavior.

The technique of snow tracking by itself cannot answer the questions asked here. A trail in the snow is a record of the animal's movements since the previous snowfall. Once a fresh snow falls, the old record is erased and a new one begins. Major movements such as den changes are typically achieved in a single trip and are likely to be obscured by intervening snowfall. In the mountains, a light snow falls almost every night during the winter. Snow tracking therefore gives a record of instantaneous behavior—activities taking place over a few days.

Before an animal's activities can be studied over an entire winter, the animal must be permanently tagged. I tagged my porcupines with radiocollars. Fortunately, the porcupines' dens were not deep enough to choke off the radio signal to the outside. I could follow animals to any den they occupied, then read the snow sign to discover current feeding behavior. Four winters of tracking revealed the following observations.

- Animals did in fact conserve feeding specializations when moving from one den to another. They even kept their feeding choices from one winter to the next. For example, Rebecca was a consistent feeder on linden, Moth on hemlock, Squirrel on sugar maple, and Finder (who shared a winter range with Squirrel) on beech.
- No animal fed on a single tree species throughout the winter. Each animal used at least two species, sometimes three. However, no more than two species were ever found on the same feeding trail.

A similar picture of conserved winter feeding choice is painted by a 1994 study of porcupines in central Massachusetts by Griesemer and colleagues. During winter 1993, the authors followed eight radiocollared porcupines. Half of these were assigned to an experimental group; the other half were assigned to a control group. The four experimental animals fed exclusively on hemlock trees. The control group fed on hemlock, red oak, and white pine. During the second week of January, the experimental animals' access to their feeding trees was blocked by installing aluminum flashing around the base of each tree. The control group was left alone. The experimental animals responded by traveling further to new feeding trees; all of these were hemlocks. The control group continued feeding as before.

The conclusion is that porcupines choose a winter diet on the basis of at least two criteria: their innate preferences and the opportunities available in the environment. Their innate preferences carry greater weight. The feeding specializations of porcupines are reminiscent of feeding specializations shown by another folivore, the three-toed sloth (*Bradypus variegatus*), studied on Barro Colorado Island, Panama, by Montgomery and Sunquist (1978). Fermenting bacteria housed in pockets off the stomach (gastric caeca) aid the sloth's digestive processes. Although the population of sloths fed on a broad range of trees (54 species), Montgomery and Sunquist found that individual animals had "modal trees"—species in which they spent the majority of their time, feeding as well as resting. No two animals with a common home range (other than mother-infant pairs) shared the same species of modal tree. Montgomery and Sunquist point out that such specialization reduces competition within the species. The role, if any, played by individual gut microflora remains to be elucidated.

Winter Range

Porcupines feed a short distance from their winter dens, as first noted by Curtis in Maine in 1941. Curtis found that 35% of all feeding occurred within 100 feet (30.5 m) of the den and that 59% occurred within 200 feet (61 m). The average winter feeding distance of the experimental animals of Griesemer and colleagues (1994) was 33.5 m. This increased to 58 m when the animals were forced to find new feeding trees.

In addition, my porcupines move relatively short distances when changing dens. Both behaviors reflect the animals' difficulty in traveling in the snow (Figure 8.6). The combination of feeding movements and den-change movements defines an animal's winter range. During the normal snowy Catskills winters, my porcupines showed a drastic decrease in movements compared

Figure 8.6. A porcupine must snowplow a trail in heavy snow. (Photo by UR)

with summer activity. For the first 3 years of radiotelemetry, their winter ranges averaged 7.4 ha (18.3 acres) and their summer ranges averaged 64.9 ha (160.4 acres). That represents an 89% decrease in movement.

But during the winter of 1984–1985, my porcupines wandered all over the mountains, testing and abandoning far-flung dens. Their winter ranges averaged 59.9 ha, a value statistically indistinguishable from normal summer range. The difference was snowfall. The weather station at Prattsville, 8 miles away, recorded only 35% of the snow failing in previous winters. For much of the winter, the mountains were bare. The following summer, New York City's Catskills reservoir system came close to drying up, the city declared a drought emergency, Mayor Edward Koch did a rain dance at the Ashokan Reservoir, and the city's water commissioner was forced to resign. Perhaps if he had been aware of the unusual winter behavior of porcupines, both he and the city might have taken more timely action.

Those dramatic porcupine responses to snow cover help resolve a discrepancy in the literature with respect to winter range. Dodge and Barnes (1975) radiotelemetered 18 porcupines in a Douglas fir–western hemlock forest in western Washington. The animals were tracked for up to 432 days each. They showed no difference between summer and winter movements. The discordant observations of Dodge and Barnes are probably best explained by local weather. Their western Washington State study site lies in the western foothills of the Cascade Range. The falling snow is relatively wet there and freezes to form a hard crust. The crusted snow offers no impediment to porcupine winter movements, unlike the deep, fluffy snow of the Catskills or central Massachusetts, in which porcupines can sink up to their bellies.

Similar factors may explain the anomalously large winter ranges observed in porcupines of a mountainous region of north-central Utah by Stricklan and colleagues (1995). Despite the report of one female moving 450 m in fresh snow, the authors report porcupines "adeptly tobogganed on crusted snows down extreme slopes to avoid capture" (Stricklan et al. 1995).

We may conclude that porcupines decrease their movements in winter not because of cold but because of snow. Their short legs and naked footpads are more useful on other substrates. In addition, porcupines in some locations may be forced to curtail winter travel because of restrictions on available den sites or food supplies.

Influence on the Forest

Because porcupines feed close to their winter dens and because they return to the same dens year after year, they can significantly influence the structure and composition of the surrounding forest. Porcupines don't affect the woods in the same way as beavers do, who may fell trees or kill them by flooding low-lying sections.

Porcupines seldom kill a tree outright. When a tree does die, the death follows from basal girdling. To measure the extent of tree killing, I tallied 812 feeding trees in current use over two winters. Only 38 (4.7%) had been girdled at the base and therefore killed. The species composition of the girdled trees was indistinguishable from that of all trees used (61% beech, 29% sugar maple, and 10% others). Girdled trees did not, on average, differ significantly in diameter from the overall tree population used for feeding (14.7 cm for girdled trees vs. 14.0 cm for all trees used). There was, however, an upper limit to the size of trees that were girdled; the largest tree girdled was 40.4 cm in diameter. Many feeding trees were larger, up to 84.4 cm for a sugar maple. Porcupines presumably avoid basal feeding on the giant trees because the thick outer bark must be laboriously removed to reach the edible inner regions.

Most porcupine feeding occurred in the crown of the tree (Figure 8.7). Of the 812 trees I have examined, 248 showed some degree of feeding (often only a small test bite) on the lower 2 feet of the trunk. When I counted only heavy feeding (removal of more than 100 cm^2 of bark or girdling of the tree), I found that 84 trees (10.3%) had suffered significant basal damage. The bulk of serious feeding took place in the upper regions of the trees, usually among the small branches of the crown.

It can be argued that porcupines are long-term agents of ecological diversification. Their winter habit of seeking out forest patches enriched with preferred trees should make it easier for rare tree species to invade such feeding areas. Unmolested by the porcupines and unsuppressed by the original dominants, the added species should contribute to the diversity and stability of the forest. The porcupine's suppression of the forest canopy should also encourage the understory, with advantages to a range of animals such as the ruffed grouse, snowshoe hare, white-tailed deer, moose, mourning warbler, yellowthroat, and a variety of other songbirds and small mammals.

Like the beaver, the porcupine is a manager of its habitat. Its management efforts are restricted to small patches of forest in the vicinity of the winter dens. Because the sites are often in rugged and inaccessible terrain, they are seldom seen by human visitors. But to a number of the permanent residents of the forest, the porcupine's patches offer food and home.

Figure 8.7. Finder in the top of a beech sapling. Most winter feeding of porcupines takes place in the canopy, where thin bark makes feeding easier. (Photo by UR)

Notes

1. Cellulose is a polymer of glucose in beta 1–4 linkage. Hemicelluloses are predominantly polymers of five-carbon sugars. Pectin is largely a galacturonic acid polymer in alpha 1–4 linkage. Starches are polymers of glucose in alpha 1–4 and alpha 1–6 linkages (Robbins 1983).

2. I have measured gut-passage time by placing wild porcupines on a 100% apple diet. Animals feeding in the wild produce bulky droppings. The low-fiber diet of apples produces much smaller droppings. The transition time between the two dropping types is approximately 48 hours. Felicetti et al. (2000), using chemical markers to measure passage time, obtained some ballpark results: 38 hours for passage of particulate foods, 57 hours for liquids.

3. $\chi^2=8.38$, df 1, $P<0.005$.

4. $\chi^2=1.9$, df 3, $P>0.5$. Not significant, assuming that only the four palatable species are subject to attack.

5. The work of R. E. Hungate (1966) has shown that, among ruminant herbivores at least, there may be a match between dietary plant species and specific gut microorganisms. For example, the protozoan *Ophryoscolex purkynei* does better on grass than on an alfalfa diet. *Diplodinium dentatum* is intolerant of clover, unlike *Epidinium* species, which prefer red clover and starchy foods. Hungate has shown, further, that certain protozoa are mutually intolerant. For example, *Polyplastron multivesiculatum* is intolerant of *Epidinium* species and *Eudiplodinium maggii.*

9 Reproduction and Maternal Care

Somewhere ahead, a porcupine is screaming. I walk through a leaf-bright October forest, tracking Rebecca. The sound starts again, close. She is up a young oak tree, a dark mass near the end of the branch, and a second porcupine sits next to her. At 5:25 p.m., the sun has already set, but the western sky glows brightly and the near-transparent October canopy affords a clear view. An autumn bite is in the air. In the Catskills, mid-September to mid-October represents the peak of the porcupine mating season. Through my binoculars I study Rebecca and her suitor. He looks coal-black, in contrast to Rebecca's more silvery hue. They seem to be equal in size.

The male advances toward Rebecca, and she retreats to the tip of the branch (Figure 9.1). Her back is turned to her would-be lover, and as he tries to reach her, she gives him little slaps with her tail. The male retreats—porcupines are as vulnerable to quills as other animals. Rebecca is howling, the waves of sound washing through the empty woods. Her suitor answers more softly. He reaches out his paw to her tail, but she slaps again.

At one point, Rebecca seems to be weakening; her tail is half-raised. The male immediately puts his head down and tries to sniff her rump. Then he elevates himself on his haunches as though to say he is ready. But Rebecca is not.

The male is enterprising. He retreats to the trunk of the oak tree, climbs up to the next branch, and heads out over Rebecca, trying to reach her from above. But the distance is too great, and he returns to his starting point.

Rebecca rebuffs him the way a buffalo rebuffs her male—she sits down. Then she gets up and turns to face him. She has retreated to the outermost limits of the branch and utters small cries as the male presses in. Surprisingly, he moves back toward the trunk to wait. I have been watching for 15 minutes. The woods have fallen absolutely still.

Rebecca's next action seems totally inappropriate. She nips off a twig and begins to feed. The red oaks are still holding their acorns, and porcupines

Figure 9.1. Rebecca (farthest on the branch) rebuffs her suitor. (Photo by UR)

have been mast-feeding for weeks. But as Rebecca feeds, the male keeps drawing closer and interrupting her. He is not hungry. When the male approaches again, Rebecca squawks and whirls to face him. She keeps trying to feed but drops her branch. The male retreats slightly and waits, rounded up and dark. Rebecca turns her back, which the male accepts as encouragement. He touches her gently on the rump. Rebecca jerks, spins around, and screams. The male shuffles all the way back to the trunk, descends to the next lower branch, and tries to find a pathway up from below. There is none. He comes back to his previous position. The light is getting dusky now, and my breath is more visible in the cold, but the porcupines look fluffed out and warm.

The male approaches again with reassuring chuckling sounds but is thrown back by a scream. At 6 p.m., as Rebecca goes on feeding, the male creeps up from behind. Rebecca turns and cries out, but her cry is softer now. In the dusky light, I can't be sure what happens next, but she appears to offer him the branch she has been feeding on. They feed on the same branch at opposite ends. Then Rebecca nips off another branch. The two feed side by side, about 2 feet apart.

Now the male approaches quietly and lowers his head to sniff the spot where Rebecca has been sitting. She continues feeding. I become aware of the moon shining through the trees above the ridge. At 6:20, the animals have stopped feeding and sit quietly next to each other, rounded out against the cold. Ten minutes later, they have become two black knots against the darkness of the trees. The moon catches a slight haze in the air and hangs like a paper lantern above them. I turn home. I expect that the persistence and gentleness of the male will prevail in the end, but at the same time, I realize I have not been admitted to a mystery. Despite years of study by a series of naturalists, many aspects of porcupine reproductive behavior still retain this quality of half-revelation.

Male–Female Interaction

Reproductive activity is both more and less important in the life of the porcupine than in the life of the human being. It is more important because an adult female like Rebecca can expect to spend 11 months a year either pregnant or lactating, year after year with no time off to rebuild body reserves. On the other hand, Rebecca is unlike humans in that she may copulate only once a year, during the 8- to 12-hour period that she is in estrus and solicits the male.

The male, who does not normally spend any time with the female, must find that tiny window of opportunity. As a female approaches her estrus, the membrane that had obstructed the vaginal orifice weakens and dissolves. The vagina begins to secrete a thick mucus, apparently serving as an olfactory cue for males. They follow the faint odor trail to its source. During the mating season, males wander more widely than in the months before. For instance, Homer's home range for July–August 1983 (before mating season) measured 19.9 ha (49 acres), but during mating season, in October–November, it grew to 100 ha (247 acres) (Figure 9.2). In fact, mating-season wanderlust may explain the larger overall home range of dominant adult males. Home ranges of subadult males do not expand at mating time. When Finder was 1 year old, in summer 1983, his July–August home range measured 13.3 ha; the October–November range actually shrank, to 9.0 ha (Figure 9.3).

As the males wander, pre-estrous females send olfactory signals. A female advertises her reproductive status by the odor of her urine as well as by the vaginal mucus. With the vaginal closure membrane gone, the mucus can mix with the urine, and a male can check her condition simply by sniffing the urine under her tree. More important, the urine spattering down through the canopy will generate an odor cloud that will alert males far away.

Most of the time, the male following an odor plume arrives too early. Although the female's urinary and vaginal secretions have started to signal, she is not yet receptive. The male therefore begins to guard her. He climbs into her tree and waits on a lower branch, sniffing periodically to check her status. Sometimes he gently nudges her from behind, forcing her forward on the branch so he can sniff the surface she has been covering.

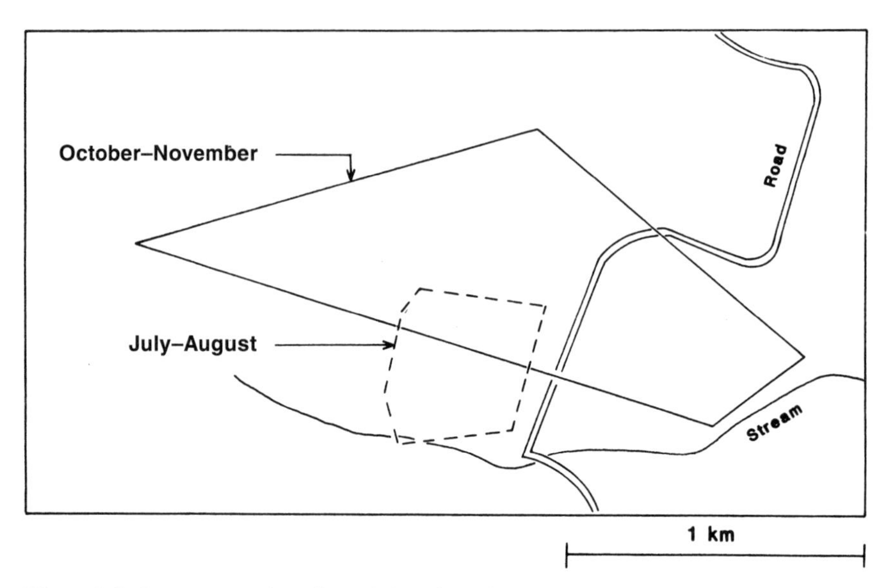

Figure 9.2. Range expansion of an adult male during breeding season.

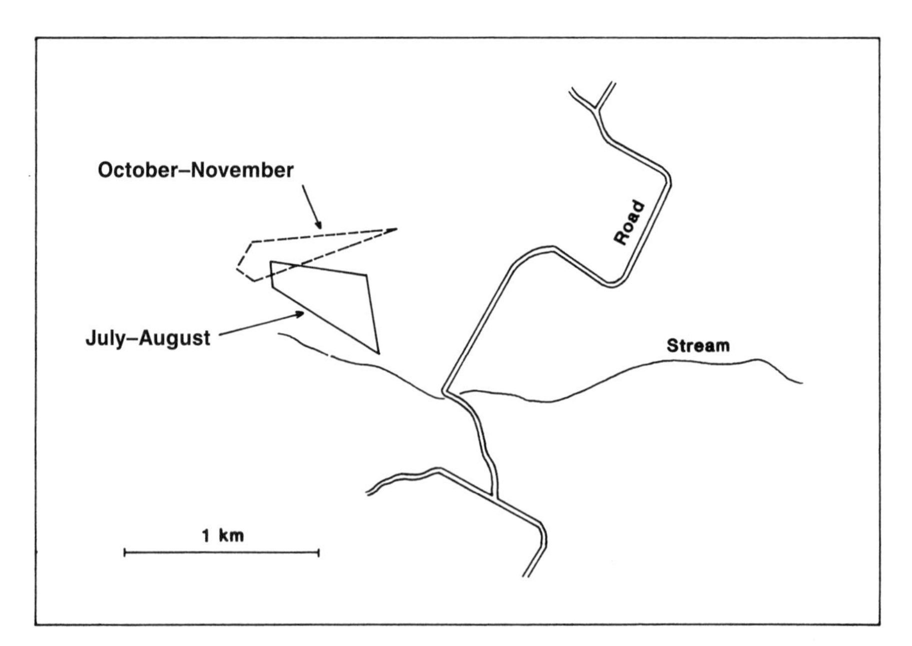

Figure 9.3. Absence of breeding-season range expansion in Finder, an immature male.

There is suggestive evidence that part of the odor signal at this stage may involve information from the perineal glands, located between the urethra and anus (Roze et al. in press) The perineal glands are present in both males and females. In males, the glands are active in individuals with fully descended testes. In females, activity increases during the fall, that is, during mating season. In both sexes, the active product is produced by a specialized bacterial flora housed in the glandular pockets and is disseminated by osmetrichial hairs that protrude from the pockets (see Figure 2.2, p. 18).

The guarding behavior may last several days, and it may involve more than one male. I got an illustration on a late-September evening as I followed Rebecca through another mating season. This time, Rebecca was up a giant yellow birch with the male Taylor, so close on the same branch they were touching. Neither animal had its quills erected, but Rebecca seemed annoyed. She gave small, sharp squawks, more than once a minute. When Rebecca turned her side to the male, he pushed his head into her lower belly. She squawked but did not budge. Taylor pulled back, then plunged his head once more into her belly, keeping it there a long time. At 7:15, Rebecca increased the frequency of her vocalizations until she was calling almost continuously but softly. At 7:35, she stopped. It was now so dark I could not see the animals. Around 8:30, a wind picked up and masked any sounds coming from above. Chilled to the fingers, I left. On my way home, I noted that the small immature male Goff was in a tree about 100 feet to the east.

When I returned the next evening, Rebecca and Taylor had moved uphill, almost to the top of the ridge. Even from a distance, I could hear squawking. I found both in a giant oak tree but on separate branches. I could not guess what Taylor had been doing to elicit the protest, but all was silence when I reached the pair. On this second day of signaling by Rebecca, two other males were relatively close. Goff had moved downhill and was now about 600 feet away. Rebecca's adult son Ram was on the other side of the ridge, about 500 feet away. For a moment I thought I had spotted another male, a large, compact shape clinging to a sapling; then, like a magician, it bent its head, raised two giant wings, and flew noiselessly away to the east—a barred owl. I hurried on to locate other porcupines, working into the night. At 1:40 a.m., from far away, I heard the high-pitched screams of porcupine, coming from Rebecca's position.

On day 3, Rebecca had moved a short distance from her previous position and was up a large oak tree, alone. But the number of her admirers had increased. A booming radio signal came from the adjacent oak tree, which held Killer, a powerful male. Killer was looking rather sorry. A second male was guarding him. Killer was as far out on a branch as he could get, his tail turned to his pursuer. I saw at least two white quillshafts embedded in the other male's muzzle. I recognized the animal. It was not Taylor but Mayday, a large male who had dropped his transmitter 4 years previously, retaining the nylon collar around his neck. The collar emphasized his muscular build and suggested a determined nature.

Under the tree I found a storm of loose quills, some commingled with tufts of hair. Some quills were kinked and toothmarked, suggesting that the victims had extracted them from their own skins. Many more quills lay under Rebecca's tree 30 feet away. A great battle between males had taken place here; the screams of the previous night had probably come from the males, and the battle was not yet finished. Long-courting Taylor had been vanquished, but Killer and Mayday remained so involved with one another that they had left Rebecca alone in her tree. Evidence showed the struggle had taken place on the ground. In many places, the ground cover had been torn up, with quills driven into leaves and twigs. Weapons had included the formidable incisor teeth; large tufts of hair had been torn out along with quills. The tufts had come from the sides and upper back, where the quill defense was less effective. I collected 1474 quills, with many more escaping my notice among dry leaves and twigs (Figure 9.4).

Rebecca's first suitor, Taylor, had fled about 200 feet away, on the other side of the ridge. He had been involved in the melee; there were seven dozen loose

Figure 9.4. The residue of battle—1474 quills lost by three males in battle over a female. (Photo by UR)

quills under his tree. They had been combed, shaken, or pulled out of his coat, not lost during primary battle. Goff and Ram were even farther away than on the previous day.

On day 4, Rebecca's mating encounter remained unresolved. She had moved about 200 feet west, taking Killer and Mayday with her. All three animals were up the same tree, a large, sloping sugar maple. Rebecca was resting on an outer branch; the two males remained preoccupied with each other and sat on a separate branch. Mayday, with foreign quills still projecting from his muzzle, was on the inside, guarding Killer. But Killer was no longer dangerously far out on the limb and showed little fear of Mayday. I found six quills under the tree; two showed toothmarks, indicating they had been pulled out of the skin by the porcupines above.

I do not know who finally inseminated Rebecca, but it is clear that male porcupines vary considerably in their reproductive success, more so than females (Tables 9.1 and 9.2). Whereas essentially every female bears one young a year, some males may produce no offspring and others may sire several. Mating success in males is positively correlated with weight (and probably age) of the male and with the size of his nonwinter territory. (Finder is excluded from the analysis in Table 9.1 because he suffered a serious fall from a tree in 1986, which appears to have left him permanently impaired. He had a high mating success in the previous year and a much larger home range.) Apparently the female's long preovulatory signaling facilitates the variation in male reproductive success. Because the males receive sufficient advance notice, all receptive females are likely to be located, and several males can assemble and either leave the scene or do battle for supremacy. The female is therefore assured of mating with the strongest, most persistent male. In fact, it is possible that a single dominant male may inseminate all the females within his home range.

Sweitzer and Berger (1996) and Sweitzer (2003) paint a similar picture concerning the Great Basin Desert. Male reproductive success was correlated with size of the home range, and dominant males showed physical evidence of battles with other males: breeding-season injuries and quill impalements. Injuries involved facial cuts and bare patches of skin. Do battles between males ever escalate to murder? Sweitzer and Berger found no instances of death in their study, which covered a maximum of 14 radiocollared males per year.

Perhaps the strangest aspect of the reproductive interaction is male urine-hosing of the female. The male approaches on his hind legs and tail, grunting in a low tone. His penis springs erect. He then becomes a urine cannon, squirting high-pressure jets of urine at the female. Everything suggests the urine is fired by ejaculation, not released by normal bladder pressure. Porcupines with everyday full bladders do not squirt their urine, do not have erections, and do

Table 9.1. Observed mating success (guarding episodes) of six males in 1987

Male	Guard episodes	Weight (kg)	Nonwinter home range (ha)
Killer	3	7.8	147.3
Jasper	3	7.9	*
Taylor	1	6.8	58.5
Finder	0	8.4	36.7
Ram	0	4.5	36.6
Goff	0	4.1	*

Average success rate (guard episodes) for males: 1.17±1.47
Correlation of mating success with weight:[a] $r=0.94$, df=3, $P<0.05$
Correlation of mating success with home range:[a] $r=0.99$, df=1, $P>0.05$
Note: * indicates home range unavailable because animal was collared in mid-season.
[a] Excluding Finder

Table 9.2. Reproductive success of three females, 1983–1988

	1983	1984	1985	1986	1987	1988
Squirrel	+	?+	+	+	+	+
Rebecca	?+	+	+	+	–	+
Moth	+	+	+	+	+	+
Average success rate for females: 0.94						

Notes: +, offspring produced; –, no offspring produced, ?+, evidence for maternity circumstantial (lactation during part of summer, heavy salt consumption); baby not captured or sighted.

not aim at females (Shadle 1946, 1951; Dathe 1963; Dodge 1967). Shadle, observing the process in caged animals, notes that the urine salvos were so powerful that drops of urine splattered on the laboratory floor 6.5 feet away. In less than a minute, a female may be thoroughly wetted from nose to tail, especially her belly and sides. The urine hosing may begin in the tree; then the pair descend to the ground for the consummation.

I witness a urine shower in the woods on a Columbus Day weekend in October 1989. I have been tracking the giant male Claw, and reach him high up the mountain in a large oak tree. Claw sits hunched up in a crook near the trunk, and he is guarding a female. Contrary to the usual pattern, the female is resting on a branch below him, head facing the trunk. As I watch, Claw descends to the female's branch. She bars his way and does not move until Claw's tail actually brushes her face. She retreats farther out on the branch, with short shrieks of protest, while the male sniffs the spot she had been resting on. Then Claw rises erect on the stout branch and sprays the female with two urine salvos. The droplets fly in horizontal sprays, glinting in the sun, and I am reminded of the Renaissance paintings of Danae and the golden shower. Everything has happened in total silence. As I wait, the sun disappears and a light sleet begins to fall. I leave the silent animals sitting about 5 feet apart on the branch, building to their next step.

This strange behavior of porcupines and other hystricomorphs has never been adequately explained. However, a very similar behavior is much better understood in a number of small rodents, including the house mouse. As long ago as 1956, Whitten demonstrated that in mice, male urine induces estrus in caged female mice; the process is now known as the Whitten effect. The nature of the chemical signal has been elucidated (Jemiolo et al. 1986; Harvey et al. 1989; Novotny 2003). The odorants are small molecules, whose concentration in the urine is vanishingly low and whose activity is greatly enhanced by urinary protein carriers. Hence, mouse urine, in contrast to that of humans, contains significant amounts of protein.

The same is true of porcupine urine; it contains approximately 0.5% protein, a significant amount. The porcupine urinary protein is an albumin (U. Roze and D. Matassov, unpublished results). Albumins are carrier proteins, which transport fatty acids in the blood. But in the urine, albumin is preadapted for the transport of other small molecules; it would not be unreasonable to guess that in male porcupines, urinary albumin carries odorants that induce estrus in the female. And, as noted in Chapter 13, porcupine urine conveys other messages as well.

A reproductive signaling function for porcupine urine fits well with the other behaviors of porcupine reproduction. During the mating season, when the female first begins signaling her approaching estrus, there is no guarantee that a male will arrive promptly. The female should go into estrus only when the victorious male is present and copulation can begin. And from the male perspective, the acceleration of estrus can enhance reproductive success by increasing the possible number of matings. At present, these ideas are speculation. They deserve experimental investigation.

Once the female has been brought to estrus, she assumes the traditional copulation posture of mammals: she elevates her hind quarters and curves her tail back over the quill rosette. If the female is insufficiently aroused, the tail may be only partially elevated and may require pushing aside by the male. He can now penetrate without fear of impalement. He rests his forepaws against the quill-free undersurface of the tail or lets them hang loosely. Dodge states, "Sexual contact is very brief, with a violent orgasm, after which the male drops back and proceeds to groom and cleanse himself. After short periods of rest, the play is pursued again until one or the other climbs a tree where he or she will scream at the other member, thus ending further mating attempts" (1967, 75).

According to Shadle (1946, 1951, 1952; Shadle et al. 1946), each sex act lasts 1–5 minutes, but repeated copulations may stretch the interaction out over several hours. In the end, the male lies exhausted, although the female may solicit additional copulations. Within hours, a half-inch-diameter vaginal plug, with a bluish white, starchy appearance, forms in the vulva. The vaginal plug, a reproductive feature of many rodents, is apparently formed by enzymatic action on the semen and is therefore a male product. Although its role in porcupines has not been studied, proposed functions in other animals include the gradual release of spermatozoa as the plug disintegrates, prevention of semen loss, and blockage of copulation by other males (Voss 1979). A further possibility in the porcupine is that it may reduce the female's ability to signal to other males.

Male Reproductive System

In addition to the strange courtship by urine, male porcupines show a number of further reproductive idiosyncrasies.

The baculum (described in Chapter 3; see Figure 9.5) keeps the penis normally retracted within the penile sheath, where it lies flexed like a partly folded jackknife (Mirand and Shadle 1953). The flaccid length is about 7.5 cm, the same as the depth of the vagina. Erection is accomplished in two ways. A rapid muscular contraction straightens out the penis, everting it beyond the penile sheath. That happens not in minutes, as in humans, but in 2–4 seconds. Further erection is accomplished by the same mechanism as in other mammals: the influx of blood into the three central chambers (the corpora cavernosa). The baculum has little to do with keeping the penis erect; its average length is only 1.2 cm (0.5 inch).

The glans (tip) of the porcupine penis is covered with low spines about 0.3 mm thick and 1 mm long (Figure 9.5). Most are conical, but some are double- or triple-crested. Mirand and Shadle, who describe the spines, report that they are made of a horny material and arranged irregularly on the surface. Undoubtedly the structures add something to the female's sensations during coitus, but it is not known whether they help induce orgasm. Such spines are common among hystricomorph rodents, for example, guinea pigs, agouti, paca, and nutria.

A bizarre accessory of the penis that Mirand and Shadle fail to describe is a small, eversible sac, the sacculus urethralis (Pocock 1922; Hooper 1961;

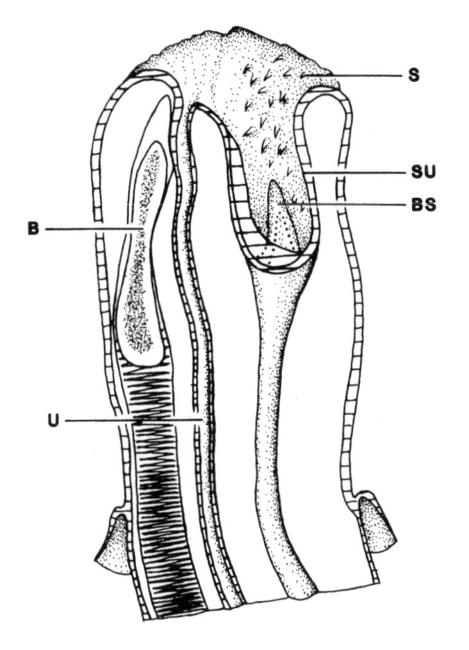

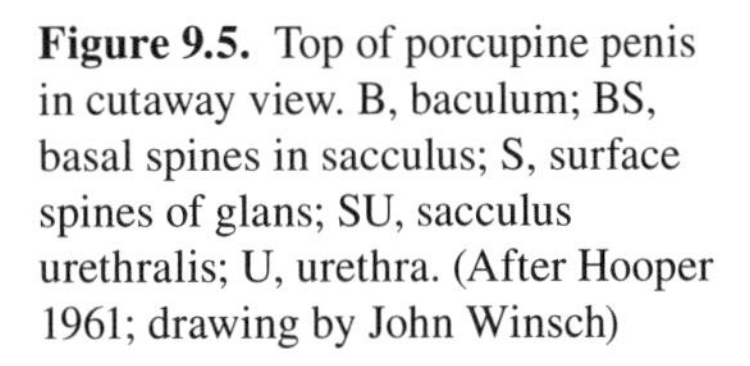
Figure 9.5. Top of porcupine penis in cutaway view. B, baculum; BS, basal spines in sacculus; S, surface spines of glans; SU, sacculus urethralis; U, urethra. (After Hooper 1961; drawing by John Winsch)

Figure 9.5). When the penis is viewed from the tip and the glans is slightly retracted, two openings appear: the urethra above (dorsally) and the sacculus urethralis below. Both openings are visible to the naked eye. The low prickles, which cover the glans, end at the urethral opening but cover the inner surface of the sacculus urethralis. At the bottom of the sacculus sit two larger spines, each about 2 mm long. During full erection, the sacculus everts, and the large spines as well as surface prickles are brought into view. Eversion of the sac is controlled by the amount of blood in the engorged penis as well as by longitudinal cords at the base of the sac. The function of the sacculus remains unknown. The male porcupine shares the extraordinary structure with other New World and Old World porcupines and with hystricomorph rodents in general. It is perhaps significant that in so many of the species with a sacculus, large spines are situated in the bottom of the sac. They should provide strong vaginal stimulation during intercourse, in addition to effects of the smaller surface spines of the glans. The sacculus may, therefore, function as a housing for penile spines too large to fit comfortably on more exposed regions of the glans.

Sperm production typically requires temperatures lower than body core temperature. For that reason, the testes of sexually mature mammals are housed in the relatively cooler environment of the scrotal sacs. The porcupine scrotal sacs are not histologically similar to those of typical mammals, but they appear to have a similar physiological function (Pocock 1922; Weir 1974). Embryonically, the mammalian testes develop inside the abdominal cavity, near the kidneys. They then descend into the scrotum through the inguinal canals. Mirand and Shadle show that in porcupines the testes descend at least twice. The first time is in the embryo, but within 8 days following birth, the testes travel back into the abdominal cavity. The second time is at the onset of sexual maturity. For my male Finder, that occurred at the age of 24 months. When examined in March 1984, at the age of 22 months, Finder's scrotal sacs were empty. In May, when Finder was 24 months old, I was amazed to find a swelling, but on one side only. The right testis had descended; its mate was still in the abdomen. In August 1984, Finder had two plump swellings around the cloaca; both testes were down. For the male Ram, one testis had descended at 25 months, with the other only half-descended.

Since the onset of his sexual maturity, I have captured Finder 13 times, at all times of year, and the testes have remained in their scrotal sacs. The same has not been true of other adult males I have examined, including Ram. Their testes have remained abdominal from roughly January to July, descending to the scrotum from August to December. My observations agree with those of Dodge in Massachusetts, who found the testes to be abdominal through most of the year, descending to the scrotal pouches during breeding season. Reasons for the variation among males are unknown, and other porcupine populations have not been studied from this perspective.

Female Reproductive Tract

Because female porcupines almost invariably produce a baby every year, despite a relatively brief annual receptive period, it has been suggested (without proof) that the act of coitus itself induces ovulation. That is the pattern in many rodents, as well as ferrets, rabbits, and others (Weir 1974). Wendell Dodge (1967) suggests another explanation for the high fecundity—females may be cyclic ovulators. If a female is not fertilized during her first ovulation period, she may recycle a month later, as in the case of human females. Dodge's hypothesis is supported by his histological examination of porcupine ovaries and by a stretched-out mating period in his New England porcupines (September to January). Among my Catskills porcupines, the peak mating period spanned mid-September to mid-October (Table 9.3).

Some strong if circumstantial evidence for recycling comes from an observation by Betty Lou Burge (1966) of a young captive female porcupine. The porcupine passed a vaginal cast on September 8, 1965, and again 28 days

Table 9.3. Mating phenology in Catskills porcupines

	Interactions observed	Rate (interactions/month)
September 1–15	3	6
September 15–October 15	22	22
October 15–31	2	4
November	4	4
December	2	2

later, on October 6. Each cast resembled a small (7.5-cm-long) translucent sausage casing. The cast represents the intact epithelial lining of the vagina, which proliferates extensively and is sloughed off intact following estrus. Burge allowed the animal free access to a male and on August 10 had observed copulation, 28 days before the passing of the first cast. The female delivered a baby on May 15, 1966, suggesting another mating at the beginning of October. Because the animals were not observed around the clock, additional copulations were possible. The interpretation is that the female went into estrus on August 10, September 8, and October 6, with copulations each time but successful fertilization only on the last try.

Pregnancy and Birth

The term of pregnancy of the porcupine is one of the longest known in the animal world. At 210 days, it approaches that of the human, whose body mass is ten times larger (Shadle 1948). Possibly a shorter term cannot be managed on the porcupine's winter diet, when females are losing weight even as their fetuses are growing. But the timing of the mating season and pregnancy accomplishes two goals. By finishing more than 90% of all matings by November, porcupines exploit the autumnal mobility of males. Once December snows have fallen, the movements of all porcupines are reduced by about 89%, and males may no longer be able to follow interesting odor trails. Second, the timing of births for early May assures mothers of a rich diet for milk production while babies find the enriched diets they need for rapid weight gain. Hence, the embryo that implants in October is implanting at the optimal time. During the winter, when essentially all females are pregnant, they conserve their energy, shelter deep in the earth, and try to maintain body weights on inadequate diets. The fetus is growing inside the mother's body but remains small. During the last month of pregnancy, in April or May, a rich spring diet suddenly becomes available, and the fetus rapidly increases in size.

At birth, the baby is well developed and weighs around 490 grams (about 1 pound). It is completely covered with fur and quills but comes into the world much like a cat, wrapped in its caul (the amnion). Inside the caul, the quills are soft. After delivery, the mother licks off and eats the caul. It contains valuable minerals and protein; its removal also keeps odor-cued predators such as foxes and weasels from locating the baby. Although the baby's quills harden during the first hour and it defends itself with vigorous tail flicks, its best defense at that time is to hide. My closest experience with birth in the wild came on May 5, 1986, when I found a porcupine baby that may have been only hours old. I had searched for babies all day. In the morning, I had radiotracked Moth to an impenetrable tangle of slash; all her behavior indicated she was hiding a baby there and not stirring far from its side. In the afternoon, higher up the mountain, I spotted an unmarked porcupine on the ground, about 100 feet ahead. The animal was conspicuous against the browns of the forest floor; the wildflower peak at that altitude would still be some weeks ahead. I outran the porcupine, captured it, and examined it. It was a 3.9-kg (8.6-pound) female, and it had recently given birth. Its vaginal closure membrane showed a small cross-shaped scar in the process of closing up again. I milked a nipple and expressed a small drop of milk. But even though I searched every fallen tree trunk and rock crevice, I could not locate her baby. I pressed on and located two more animals.

My last visit of the day was with Rebecca. She had descended to the bottom of the valley, where cutleaf toothwort bloomed in green islands among persisting browns. I reached her just as she was ascending a medium-size red maple. That meant she had been spending time on the ground, a suspicious behavior. She climbed slowly and stopped about 10 feet off the ground at the first small branch. There she remained clinging vertically to the tree trunk. She had to have a baby nearby. Almost immediately, I saw it. Instead of hiding in some crevice or shadow, it lay in plain view, pressed against the exposed root of an adjacent beech tree, a small black mass against the gray of the bark. It was the smallest porcupine I had ever seen. It did not run when I reached for it. I gathered it inside a paper lunch bag and weighed it—450 grams on my spring scale, almost exactly 1 pound. Then I examined it more closely. The head seemed out of proportion, taking up almost half the body. At first it watched me through eyes bright and open under the long eyelashes, then through eyelids closed down to a squint, as though hiding. Its incisors were fully erupted, but they were as narrow as little needles and totally white, lacking the orange enamel of adults. As I held it, it began to chatter its teeth in the traditional warning of porcupines, but the teeth did not make contact and there was only a vibrating silence. It also made small slaps with its tail, shedding diminutive all-black quills that looked like tiny splinters and measured only 5 mm long. The baby was entirely black; none of the quills carried the white contrast coloration. I smelled it up close. It was odorless, lacking the pungent signature of the adult porcupine. Pink skin gleamed through sparse belly fur. The sex was female. The footpads were tiny versions of the adult's. From its navel, a shriveled umbilical cord still hung. I estimated its age at about 1 day and wondered whether Rebecca's halfhearted climb was due to exhaustion or to maternal conflict. During the entire episode, she watched intently but made no sound and did not move to rescue her baby.

I put the baby back at the base of its beech tree, and it made a pathetic attempt to hide, pushing its head partially under a root but leaving the rest of the body exposed. It lay still, then opened one eye to see if I was gone. I went, after making one more observation. Nowhere in the vicinity could I find an accumulation of porcupine droppings such as adults leave when they lie in one spot for a few hours. In a number of mammals, the mother stimulates defecation of young offspring by licking under the tail, then eating the feces, thus robbing a predator of a valuable cue. I have found the same cleanliness around all baby porcupines discovered in the woods. Only when the baby is old enough to climb do tiny droppings begin to accumulate under the tree.

At the time of the discovery, I did not have a chance to examine the mother. Two weeks later, I weighed her at 5.5 kg (12.1 pounds). Assuming that was her weight at the time of delivery, the baby represented 8.2% of the mother's weight. For comparison, a 7.25-pound human baby delivered by a 125-pound mother represents 5.8% of the mother's weight. Because of such a high maternal investment, twin births appear to be at least as rare among porcupines as they are among humans. Of 35 porcupine babies I have encountered over the years, all have been singlets. Shadle (1950) states that of several thousand births reported, not a single twin birth has been found. Neither Dodge (personal communication) nor Sweitzer (2003) has ever observed a twin birth.

Rebecca's baby grew rapidly. The baby essentially doubled her weight in 2 weeks. Later in the summer, the weight gain slowed but nevertheless remained well above the rate shown by her mother (Table 9.4).

Lactation

One reason for the differential mother–infant weight gain must be that the mother is feeding on tree leaves while the baby is feeding on tree leaves plus milk. Even though the porcupine bears only one offspring at a time, it has four nipples, one pair under the armpits and the other pair abdominal. Over

Table 9.4. Weight changes in mother and offspring

Date	Weight Rebecca (kg)	Weight baby (kg)
February 23	5.6	—
May 5	—	0.45
May 20	5.5	0.85
July 6	5.65	1.85
July 26	5.95	—
August 3	—	2.15

the years, I have discovered two females with an odd nipple count—each carried five. And two other females, a mother-daughter pair, had only two nipples each. According to Shadle (1950), the baby takes turns nursing at all four nipples. The mother squats on her haunches and tail, allowing the baby to nose about her belly (which is quill-free). When the baby locates a nipple, it tugs at it for a few seconds, then goes on to another, and another, and another, and back again to the first. A baby eventually establishes a fixed sequence of nipple usage, with only an occasional variation. A nursing baby accompanies itself with soft cooing, grunting, and smacking of lips and often works its small paws through the hairs around the nipples. Having exhausted the milk supply, it wanders off to play or sleep.

Shadle observed captive animals. I have observed nursing in the wild by surprising a mother and baby at night. The mother was Squirrel; I reached her at 2:10 a.m. on an August night. Her baby was 15 weeks old, weighed 2.5 kg, and was able to climb large trees and travel long distances with her mother. The two had been resting on the base of a J-shaped cherry tree, whose base was only 30 degrees off the horizontal. As I arrived, mother and daughter scooted up the tree, with Squirrel in the lead. Squirrel had chosen her nursing station well; thick surrounding vegetation blocked cold night breezes, and a shadbush feeding tree was available nearby. At 4:30 a.m., a squeaking and mewing broke out in the treetop. Then I heard the scraping sounds of porcupine descent and saw two dark masses descend to the lower, sloping portion of the trunk where the animals could rest comfortably. If I had stood up, I could have touched them. Now nursing began, and it was a musical affair. There is evidence that in some mammals the baby's vocalizations during suckling stimulate milk let-down (Tindal and Knaggs 1970). Both Squirrel and her baby were vocalizing continually, the baby in a higher register, the mother answering in contralto. The sounds had the incongruous quality of metal zippers riding up and down. At one point, the baby made a soft "mmmm" sound, which I associate with mother-searching. At no point did the vocalizations take on the sharper edge of anger. The music continued uninterrupted for 30 minutes; then I stood up and sent both animals fleeing up the tree. After a few minutes, the nursing sounds resumed in the upper branches and then died off. A short time later, thrushes began their morning chorus.

Throughout most of the period of her lactation, a mother gives little indication of her condition. For a week or so following delivery, the areolae around the nipples are pink and swollen, and milk can be obtained by gently pulling the nipple, as one would pull the teat of a cow. Porcupine milk, even then, is produced sparingly. It appears as a glistening white drop at the tip of the nipple, not as a stream. But within 2 weeks the swelling subsides, and milk can no longer be obtained manually, but that does not mean the porcupine has stopped lactating. To test for lactation at that time, I use the oxytocin method. Oxytocin, a hormone released by the posterior pituitary, lets down milk when the baby begins to suckle (Tindal and Knaggs 1970). It can release only preformed milk and does not by itself stimulate milk synthesis. The test is performed by injecting a small amount of the hormone in a leg or arm muscle. Within a minute the areolae swell as lactiferous ducts send down milk. When the nipple is gently pulled, milk floods the tip.

My use of the oxytocin method showed that porcupines in the wild spend a surprisingly long time lactating. Unfortunately, it is difficult to trap the female just after birth (to fix the starting point) and just before dispersal of the offspring (to fix the end point). Table 9.5 therefore gives figures both for observed (minimal) periods of lactation and estimated (most probable true) periods. The estimated days of lactation are obtained by extrapolating back to the probable date of birth and forward to the last observed contact between mother and baby.

Since Acidie's baby, when first captured on June 22, 1985, already weighed 1 kg, it had to be around 3 weeks old. Hence, the term of lactation for Acidie would be around 113 days. Likewise, Rebecca's baby probably stayed with its mother at least another 4 weeks (since the other babies did so); hence, the estimated term is 110 days. Squirrel's 1986 baby was at least 2 weeks old when first discovered. In addition, Squirrel was observed in the same tree with her baby as late as September 13, when she was presumably still lactating. Hence, her estimated term is 112 days. Squirrel's 1987 baby was approximately 1 week old at time of first capture. It stayed with its mother until September 26, when it dispersed. Thus 30 days must be added to the period of confirmed lactation in 1987. Moth's 1988 baby was approximately 1 week old when first captured. The best estimate for the five mother-offspring pairs listed is therefore 127 days of lactation.

This figure is surprisingly close to a large 1993 study by Sweitzer and Holcombe from the Great Basin Desert of Nevada. Using an estimation

Table 9.5. Observed and estimated periods of lactation in four females

Mother	Birth of baby or first observed lactation	Weight baby (kg)	Last observed lactation	Observed lactation (days)	Estimated lactation (days)
Rebecca	May 5, 1986	0.45	July 26, 1986	82	110
Squirrel	May 9, 1987	0.50	September 3, 1987	116	146
Squirrel	June 7, 1986	0.85	August 26, 1986	81	112
Acidie	June 22, 1986	1.00	September 22, 1985	92	113
Moth	May 15, 1988	0.60	October 8, 1988	146	153
			Average	103	127

procedure similar to mine, the authors report an average lactation period of 126±4 days.

That is a staggering figure. The average period of gestation for the porcupine, as established by Albert Shadle, is 210 days. When that is added to 127 days of lactation, the female is spending 337 days a year (11 months) either being pregnant or nursing a baby. Shadle, who kept a porcupine colony at the vivarium at the University of Buffalo and allowed the animals to breed, found mothers nursing their offspring for 90–150 days. With the latter figure, a female would spend 360 days per year in pregnancy and lactation. Shadle believed that such long periods of lactation would not be observed in nature. Yet the behavior of wild porcupines, whether from the Catskills or the Great Basin Desert, has been little different from the captives'. Ecologists point out that a long period of maternal care is needed to compensate for a poor resource base. A folivorous (leaf-feeding) diet is among the poorest available in nature; porcupines are therefore following ecological prediction in their extended maternal care.

What is surprising is that the wild females who spend 337 days a year in pregnancy and lactation do so year after year, with essentially no years off for recovery. A survey of three females over 6 years shows a minimum of 15 babies produced (see Table 9.2). The data include two question marks. No baby could be found for Rebecca in 1983, yet that May she was behaving as a typical young mother, hovering around a hollow den tree and seeking salt. She probably produced a baby and lost it at an early stage. Squirrel's 1984 maternity status is more questionable. No baby was ever found, and she did not salt-chew extensively, but she behaved suspiciously in mid-June. Unfortunately, I was not using the oxytocin test in 1983 or 1984 to establish lactation status. Rebecca did not produce a surviving baby in 1987, as shown by direct observation and confirmed by the oxytocin test. That year her weight surpassed Squirrel's for the first time, demonstrating the cost of lactation. Squirrel was heavier than Rebecca every year from 1983 through 1986 and again in 1988. Moth was the lightest of the three in every year of observation.

The 89% annual rate of pregnancy calculated from Table 9.2 (18 porcupine-years of study) is again remarkably similar to a 92% pregnancy rate in adult Great Basin porcupines (50 porcupine-years of study; Sweitzer and Holcombe 1993).

Mother–Infant Relationships

For the first 6 weeks of life, a baby porcupine is too weak to travel long distances, and the mother does not move far from its side. Young babies cannot climb trees above sapling size because for the small animals the trunk of a large tree is essentially a flat wall. As a result mother and baby meet only at night. During the day, when the mother sleeps in some large resting tree, the baby is hidden on the ground below.

I have found young babies installed in every sort of hiding place. Squirrel's 1983 baby was hidden in a rock crevice about 100 feet downhill from the mother's resting tree (Figure 9.6). Rebecca's 1986 baby, at the age of 3 weeks, lay pressed into a small cavity under a fallen tree trunk, its dark back an extension of the shadow. Rebecca was high up in a tree about 25 feet away. Acidie's 1985 baby lay in plain view on top of a rotten log, looking like a dark knot on the surface. And Moth's 1986 baby remained inaccessible for weeks inside a jumble of tree-crown slash until its curiosity betrayed it when it climbed out to explore. But the most common hiding place by far was the hollow base of a tree. Mothers often spent the day resting in the tops of such hollowed trees. By shining a flashlight inside the recess, I often discovered the baby within. That was how I first found Rebecca's 1985 baby, Moth's 1984 baby, Squirrel's 3-week-old baby in 1983, and others through the years.

Tryon (1947), who delivered a baby porcupine by cesarean section and raised it to the age of 2 years, comments that the baby's first instinct is to follow any moving object. Presumably the babies sleeping in the hollow base of their mother's resting tree followed the mother to the tree, found themselves unable to climb, and hid themselves in the inviting dark recess.

When mothers forage in treetops at night, the babies also follow to the feeding tree, then lie down at the bottom to wait. Baby and mother always come together at night; their resting positions during the day show increasing separation as the baby grows older, travels greater distances, and chooses different resting trees for daytime sleep. At night, it is the mother who travels to the baby, sometimes from half a mile away. The mother moves rapidly to the rendezvous, suggesting that she is following landmarks and not an odor trail. If mother and baby are foraging in separate trees at night, the mother seems to have a good idea of just where the baby is.

Figure 9.6. Squirrel's baby hidden under a root tangle. (Photo by UR)

Mothers never defend their babies, even when capture occurs in sight of the mother. One reason may be that animals like Rebecca, who wear radiocollars, have been recaptured so many times over the years that they may feel some loss of confidence. Naive porcupines may act more aggressively; I have been charged twice by naive porcupines disturbed in their dens. Both charges stopped short. Also, Catskills porcupines may be timid because they are generally smaller than porcupines in other parts of their range. John Bryant of the University of Alaska tells me that Alaskan porcupines, who are larger than those of the Catskills, frequently charge when provoked. Finally, the porcupine baby itself is not defenseless. It is born fully quilled and with defensive reactions well developed. Its long residency in the uterus represents a form of maternal protection. By mid-summer, its rapid growth has probably made it as effective as its mother against small predators.

Separation

With nightly contact with its mother, and with gradually increasing travel together, a baby's independence grows (Figure 9.7). At 3 months, it may begin to spend occasional nights a short distance from its mother. On the following night, the two are found together again. Figure 9.7 shows the growing

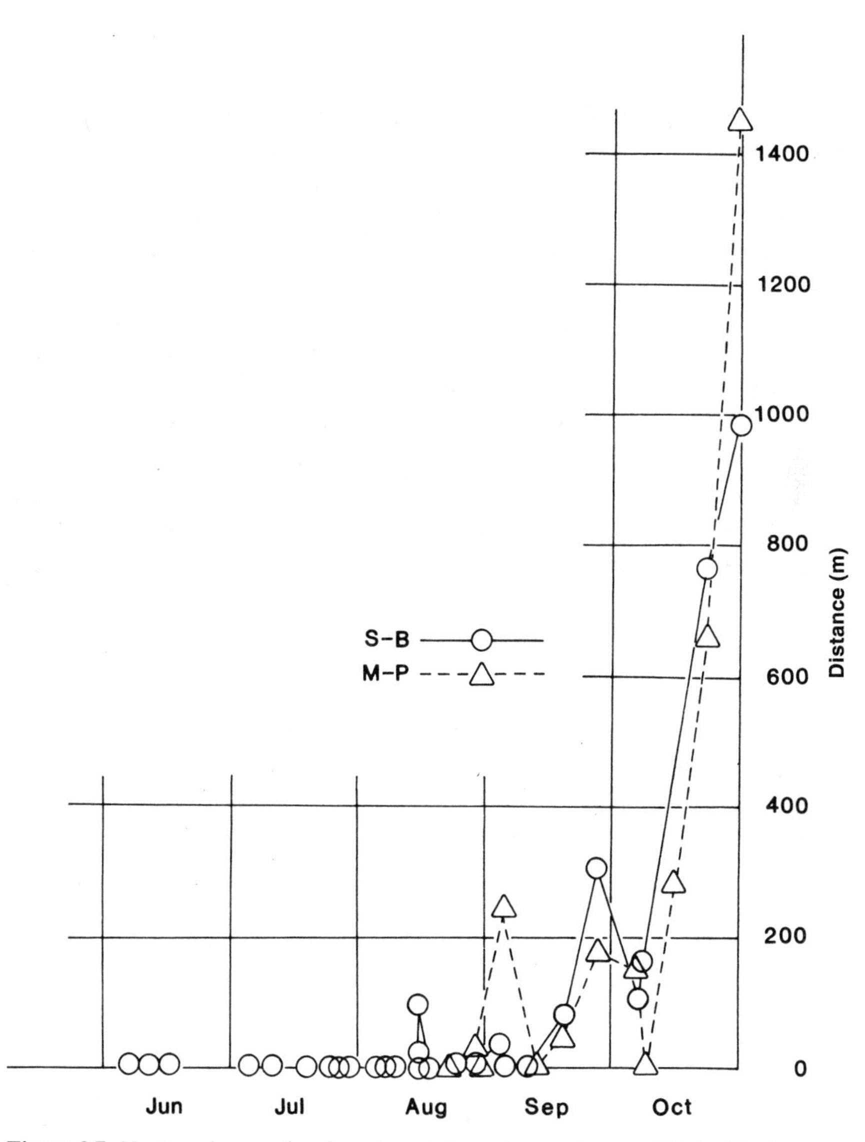

Figure 9.7. Nocturnal separation, in meters, between two mothers and their offspring. Both babies showed increasingly distant separations as the summer progressed, with permanent loss of contact in mid-October. The female offspring B left her natal range permanently. The male offspring P entered a winter den on his natal range. B, Bus; M, Moth; P, Pleiades; S, Squirrel

independence of Bus, Squirrel's female offspring of 1983, and Pleiades, Moth's male offspring of the same year. I established nocturnal separations of the animals by means of a measured-thread device. (Thread unwinding from a spool activated an odometer.) After a while, the forest floor was covered by a giant spiderweb of nylon threads. The data show that mothers and offspring seldom separated through August and the first 2 weeks of September. During subsequent weeks, which coincided with the peak of the mating season, mothers and infants were rarely found together but remained relatively close to one another. Long-distance separation occurred at the end of October.

An indicated separation for a night does not prove that the mother never met her baby that night because I did not monitor the two for all the hours of darkness. The question is important because I found that a mother who does not provide milk for 3 nights in a row stops lactating.

During their time with their mothers, Bus and Pleiades learned the identity and location of the rare food trees. They learned the locations of dens and other shelters and hiding places. After separation, their further learning would come through their own efforts. They set out energetically to survive the biggest test of their lives—their first winter alone.

10 Social Structure

A porcupine leads a solitary existence throughout much of its life. Nevertheless, it participates in complex and predictable social interactions, as these two examples of male-male meetings illustrate.

Strangers

The first story begins after midnight in July 1983, as my daughter Rachel and I are observing Rebecca's feeding activities in a spreading wild apple tree above our wellspring. The tree stands in a small clearing; unbroken forest begins just to the north and west. We are prepared to listen to feeding sounds from the apple tree. Instead, as soon as we sit down, we hear a distant, high-pitched screaming from woods to the west. Any sustained disturbance in the night forest is strange, and I wonder what animal might scream so persistently. Suddenly I know. These are the howls of a porcupine. We gather our gear at once and set out in the direction of the tumult. When screams are ringing in our ears, we turn off the trail and look up. A drama confronts us.

The lower branch of a small ash tree holds two porcupines, about 12 feet off the ground. The first, the size of an adult male, is about halfway out on a branch, facing a second porcupine, which is hanging upside down at the end of the branch, in imminent danger of being thrown to the ground. Its quills are maximally erected, its tail points toward its adversary. This is the animal that is screaming without interruption. I see that it is wearing a radiocollar, so I quickly tune my receiver. The distressed animal is Woodruff, a giant male I collared 2 months previously, when he weighed an impressive 7.7 kg (17 pounds).

Under the combined illumination of two bright headlamps, the dominating porcupine breaks off his attack and retreats to the trunk of the tree. He looks muscular, and to my eye triumphant, but I see in my binoculars that Woodruff has inflicted some pain. The attacker's muzzle is filled with quills; many appear to be broken off halfway. After a short pause, he begins to climb up the trunk, his back partially erect. Woodruff remains hanging and continues to scream, with interspersions of tooth-chattering, for about 10 minutes. Then the cries grow less intense until he falls as silent as his opponent. He rights himself on the branch and moves back toward the trunk. As he sits hunched, I see the marks of battle. A large white quill has been driven into his upper lip. I also see three white quills in the chest and one in the arm. Woodruff remains at the base of the branch, reluctant to climb after his opponent or descend toward the two humans.

There are no loose quills under the tree to indicate previous thrashing on the ground and no niptwigs to indicate feeding. I find several dense, fat droppings of the type that Woodruff usually produces, evidence that he has spent several hours in the tree. The fight tree is about a half mile east of Woodruff's previous position and perhaps a quarter mile out of his normal range. I surmise that he left his home range either to visit a salt source or to feed in an apple tree and was caught by a resident male. The experience must have been terrifying for him; in more than 600 captures, I have never heard a porcupine scream with that intensity.

Woodruff never again returns to this section of the woods. Next day, he has retreated to his usual haunts high up the mountain. As I study him in my binoculars, I see he has already removed all visible quills. He stays in the high mountains for the rest of his life, until killed by a hunter the following winter. On autopsy, I find four large quills buried in the body fascia, a permanent residue of his July-night encounter.

Woodruff's battle with the uncollared porcupine, although spectacular, fits our understanding of much of male mammalian behavior (including human male behavior). It also fits other reports of porcupine intermale aggression. Wendell Dodge has stated that male porcupines in captivity may kill other males. And Sweitzer and Berger (1996) discuss porcupine intermale aggression as evidenced by quill impalements and injuries.

Friends

The second example of male-male interaction follows an entirely different script. This story begins well after midnight in August 1986 as I capture a giant porcupine at the salthouse. His eartag identifies him as Killer, a male I tagged the previous summer. I want to take him to the city for a feeding study. By the time I have made a physical examination and placed him in a cage, it is 3:30 a.m. I want to drive to the city at once, to take advantage of the cool of the night. When I pass once more by the salthouse to lock up, I find another porcupine chewing the salt sticks on the east wall. It is much smaller than Killer and offers little resistance as I grab it. I make a quick physical exam. It is a young male, with eartag no. 58 in its right ear. I have captured Ram, Rebecca's baby of the previous year.

The capture of the 15-month-old male crystallizes something in my mind. Two radiotagged babies had dispersed out of their natal territories at the end of their first summer of life; both were females. Studies of male dispersal have suffered from experimental glitches, but they suggest that males do not disperse. I have captured Ram inside his natal territory.

I decide to include him in the feeding study. I put him in a separate cage in the back of the station wagon and drive off at once, arriving in New York at 6:30, in full daylight but before sunrise. Of the two males, the smaller one is more alert and aggressive, clinging to the sides of his cage and giving small warning screams when I come close. I install them in adjacent cages in an ivy-covered recess behind our house. Here they will be shielded from the August sun but get some cross-breeze. I fall into an exhausted sleep.

When I wake at noon and check the porcupines, I am shocked to find Ram's cage open, the animal gone. The door-retaining spring lies on the ground, worked loose by the captive. I search all nearby trees and find nothing. Where would a porcupine go in New York City? I walk along the railroad tracks to the Ravine, a wild stretch of old willows and box elder, and try to view it through the eyes of a Catskills porcupine, used to resting in cool forests and feeding on linden, aspen, and apple. I don't recognize a single familiar feeding species. What can Ram find here to relieve his hunger? What if he encounters a family dog? New York City dog owners don't expect their pets to return home with a muzzle full of quills. What if a child tries to play with him? What if he gets in the way of a car? I search a long time but return home empty-handed. I feel devastated. I have sentenced the small animal to death.

After nightfall, my neighbors Don and Joan Curran tell me they saw a porcupine moving near a fence on Depew Avenue. They clapped their hands, and the animal ran to the other side, under Mrs. Turner's pine trees. "But of course," Don adds, "we know there are no porcupines living in New York, so it had to be something else, perhaps a raccoon." A small feeling of relief comes over me. The animal is still in the neighborhood. I plan to resume my search the following day, but that proves unnecessary.

During my search, I have left Killer in his securely fastened cage outside the house, with a handful of apples for food. I check him for the last time at 12:30 a.m., only to find two porcupines in the beam of my flashlight. Killer is noisily demolishing an apple inside the cage, and the little porcupine is clinging to the wire mesh outside, as though drawn by the big male's company. I run into the house, grab my rubber gloves, and have Ram captive again in a minute. His coat is dusty, suggesting he has spent time in some ground-level hideout, not up a tree as I had expected. His mouth is dry, and he is hungry when I offer apples. The sudden reversal of fortunes is stunning, but so is the behavior of the two males. Far from trying to kill each other, they have reacted as friends. I keep them for 10 days in adjacent cages; their behavior never deteriorates. Is it possible the animals have known each other before?

I conclude they have. At the end of my feeding study, I equip both males with radiocollars and release them in the mountains at the site of initial capture. My tracking during subsequent months reveals that they not only inhabit broadly overlapping territories but sometimes occupy the same tree. At times, I find them using the same tree on alternate weeks or resting in trees near each other. At the end of September, I find both up a large oak tree near the top of the ridge; Rebecca is in a tree directly below the pair. On November 29, Killer is up a leaning linden tree and Ram inside its hollow base. There is never any hint of friction between the two. They separate only in the winter, occupying different dens or hemlock trees.

Male Territoriality

At the time of his urban adventure, Ram was still a juvenile with undescended testes. Killer's and Ram's territories overlapped the territory of Taylor, an adult male almost as large as Killer (see Figure 10.1). Killer and Taylor have at times rested peacefully in adjacent trees and in the same tree in successive weeks, but they have never, in my experience, occupied the same tree at the same time. Still other males shared the territory of Ram, Killer, and Taylor. An unidentified male was guarding Rebecca in a ridgetop sugar maple during the 1986 mating season while Killer and Ram remained a short distance away (in opposite directions) and Taylor was much farther away. Another male, a juvenile, was captured in 1987. There may be still others.

Generalizations on Porcupine Territoriality

The following generalizations concerning porcupine territoriality have emerged from my studies.

- Females exclude other females from their territories; male ranges may overlap extensively with those of other males as well as females. A single male may overlap the territories of up to 5 females. This agrees with

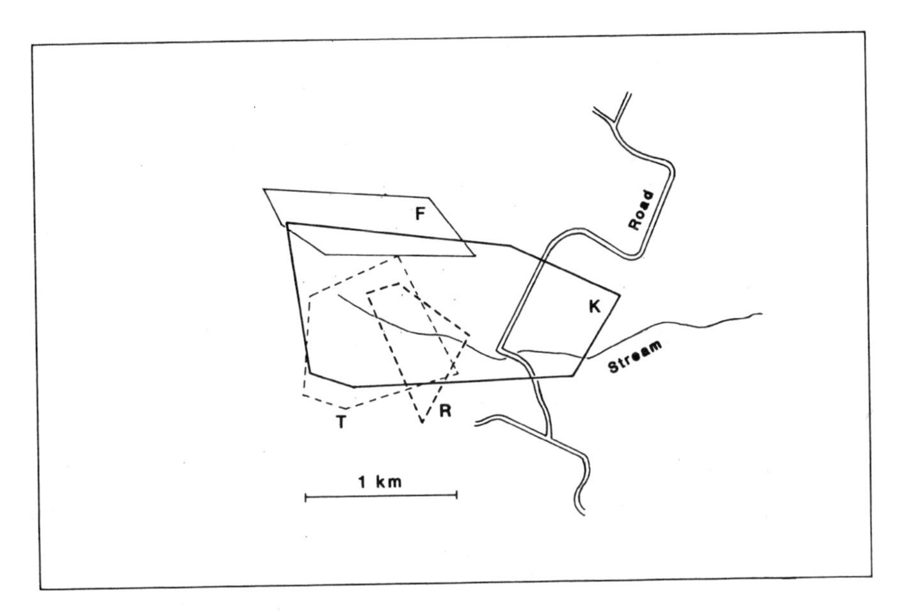

Figure 10.1. Nonwinter ranges of males Killer (K), Taylor (T), Ram (R), and Finder (F). In contrast to female ranges, male ranges overlap heavily and show great variation in size.

den-sharing patterns during the winter (Chapter 7), when identified den-sharers consisted of male-female pairs.

- Certain males show little overlap with others. For example, Finder was never found close to Killer, Ram, or Taylor, though his home range partly overlapped Killer's.
- Females inhabit territories of roughly equal size. Male home ranges may vary considerably in size, with the smallest ranges occupied by juveniles and subordinates and the largest ones occupied by dominants. Males expand their natal territories as they mature.
- Females disperse as juveniles. All adult females holding established territories were born elsewhere. By contrast, males remain in their natal territories.

Hypotheses

The data available are consistent with the following hypotheses.

- Male dominance hierarchies determine access to estrous females (Chapter 9). Competition for females has been observed directly and is also suggested by the higher body weights of males compared with females.

Males from different local groups may be intolerant of one another, as were Woodruff and his antagonist. It is not known whether such antagonism is a group behavior or is limited to dominant males.

- Although porcupines are largely solitary, males interact amicably within loose associations (local groups). Members of a male association cooperate because they are genetically related. They may share a common mother, a common father, or both. As yet no direct evidence, such as DNA microsatellite data, supports this hypothesis. The failure of young males to disperse and the stability of adult female territories over many years, however, suggest that local males share maternal genes. Whether they also share paternal genes depends on the time span a dominant male can control his extended harem.
- Conversely, adult females behave agonistically toward one another because they are unrelated and must compete for scarce resources. For that reason, females are more territorial with one another; their territories are relatively small and defensible. Female intolerance extends even to mother-daughter pairs once daughters become independent and mothers enter their fall mating period.

Those hypotheses are still preliminary; their confirmation will require additional data. Next, I first present the existing data on which my conclusions are based and then discuss the implications of my hypotheses and compare porcupine social structure with that of some other mammals.

Female Territoriality

Female porcupines show strong site fidelity. Year after year, each can be found foraging in her own core section of the woods, resting in established resting trees and feeding in the same lindens, apples, and aspens. Year-to-year changes in territorial boundaries tend to be minor compared with changes observed in males. But even males remain true to a core area, returning year after year.

Figures 10.2, 10.3, and 10.4 illustrate changes in the nonwinter territories of the females Rebecca, Squirrel, and Moth. Neighboring ranges overlapped extensively in 1983, even though the core areas remained exclusive. Fewer data points are available for Moth than for the other two females because she was radiocollared in midsummer.

Females defend their core areas against other females. It is difficult to catch them actually battling in the woods, though I have found circumstantial evidence that they do. Because defense of territory is a form of competition, we

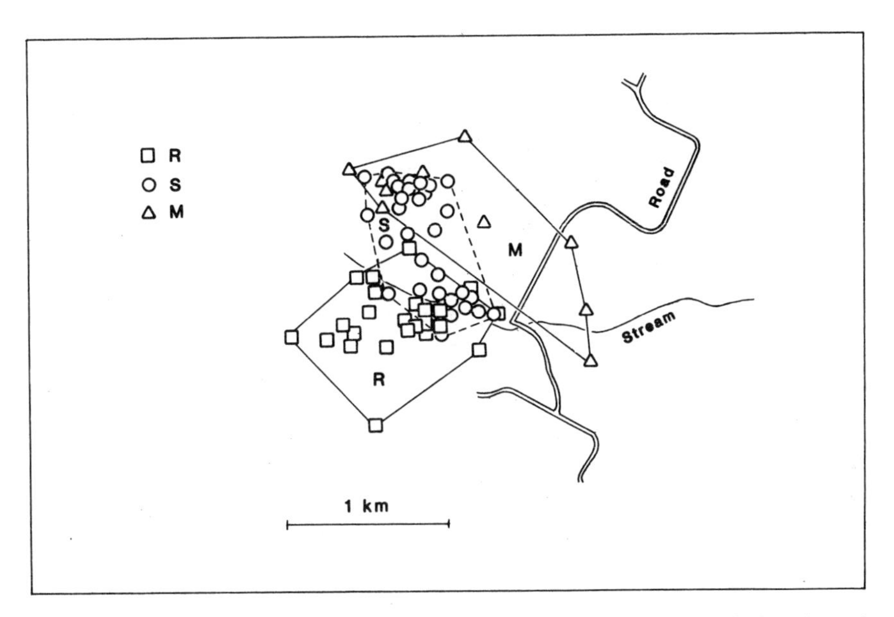

Figure 10.2. Female nonwinter ranges, 1983. Although neighboring territories overlapped extensively, the core areas showed little overlap. M, Moth; R, Rebecca; S, Squirrel

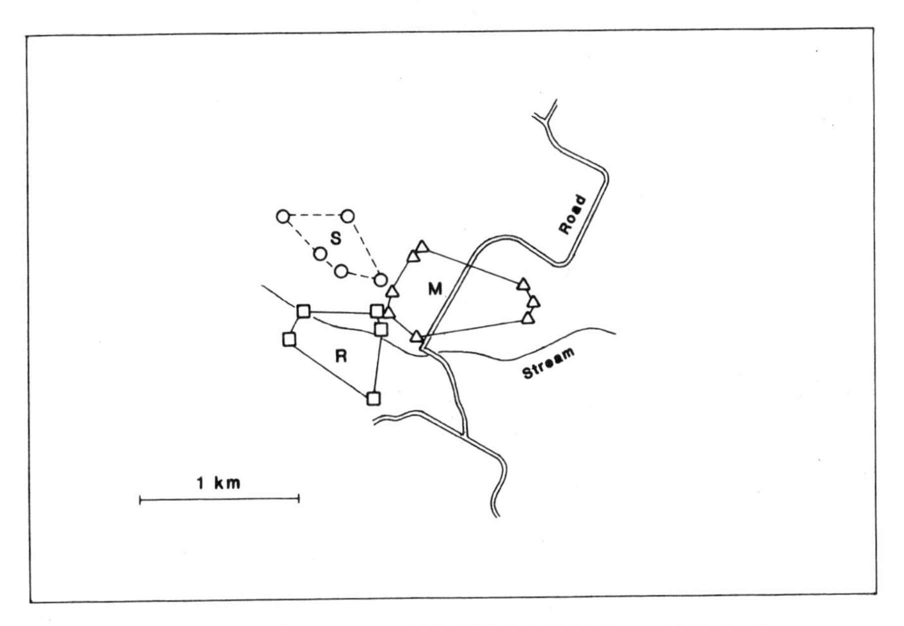

Figure 10.3. Female nonwinter ranges, 1985. M, Moth; R, Rebecca; S, Squirrel

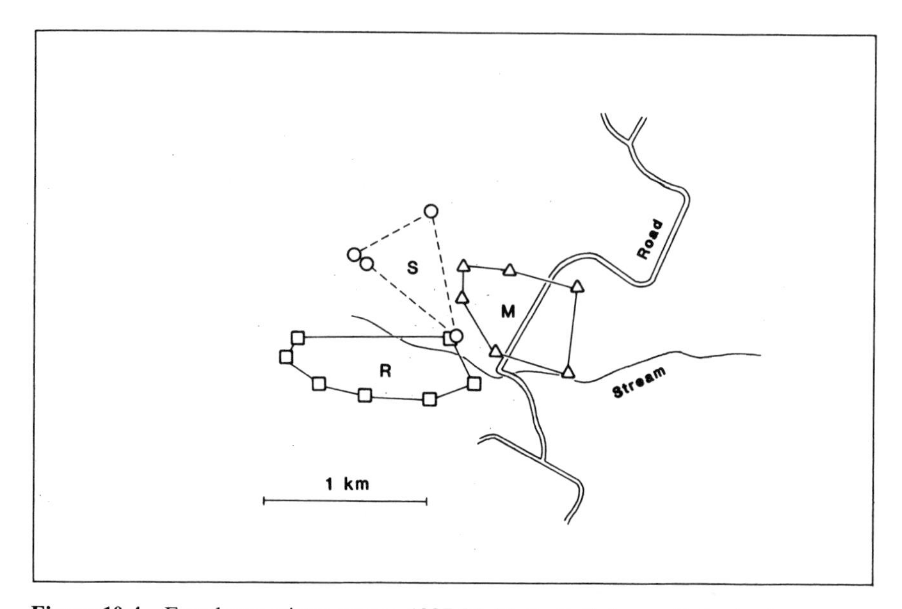

Figure 10.4. Female nonwinter ranges, 1987. M, Moth; R, Rebecca; S, Squirrel

might ask what happens to the size of female territories if the level of competition is stepped up by an increase in population density. Without being aware of it, I was studying just such an experiment in 1983. A scabies epidemic ran through my study population and reduced its density. The epidemic burned itself out in 1984, and the porcupine population rebounded. The change in population density was accompanied by complementary changes in size of female territories. In 1983, males and females had roughly equal home ranges (65.2 ha for females; 71.6 ha for males). Over the following 5 years, the average female home range contracted to 23.2 ha; the average male home range remained constant (Table 10.1). During the same period, population density approximately tripled (Table 10.2).

Another way in which the territorial instinct is manifested is through homing behavior. Animals removed from their territories will return when given the chance, as did the adult female Gro. When I found her in early April 1983, she was so weak and unresponsive that she seemed near death. I picked her up without resistance and carried her to the cabin without anesthetizing her. She had lost 30% of her summer weight—a dangerous condition—and she was overrun by lice. But on a diet of apples and chow in the city, she regained her strength and weight. In May, I returned her to Lexington and released her into a countryside of wildflowers, bursting buds, and birdsong. I made one mistake. Instead of returning her to the site of capture, on top of the mountain, I released her behind our cabin, far to the east of her normal territory and in an

Table 10.1. Male and female nonwinter home ranges, 1983–1988 (ha)

	1983	1984	1985	1986	1987	1988
			Females			
Moth	78.4	63.9	32.9	40.0	35.9	19.0
Rebecca	67.0	60.6	21.4	21.6	17.9	21.6
Squirrel	50.1	71.3	12.6	21.6	30.8	29.1
Average	65.2	65.3	22.3	27.7	28.2	23.2
			Males			
Imp	77.5					
Homer		109.5	30.3			
Topmost		60.0[a]				
Finder	65.6[a]	70.1	31.9	70.2	36.7	
Killer				86.2	147.3	84.4
Taylor				107.1	58.5	
Ram				28.0[a]	36.6	
Jasper						75.6
Average	71.6	79.8	31.1	72.9	69.8	80.0

Notes: Home ranges calculated by minimum convex polygon method. None of the male-female differences are significant by *t*-test until 1987. In 1988, when both males were dominants, male-female differences were significantly different ($t=1.1$, df=3, $P<0.01$). Female territory sizes differ significantly between years: $F_{5,12}=13.62$, $P<0.001$ (ANOVA). Male territory sizes do not differ significantly between years (ANOVA).

[a] Subadult

Table 10.2. Population density and female nonwinter territory size, 1983–1988

Year	Unique captures/year	Captures/ 100 days	Female territory (ha)
1983	14	12.1	65.2
1984	18	16.4	65.3
1985	26	25.7	22.3
1986	27	30.7	27.8
1987	26	30.6	28.2
1988	39	37.1	23.2

Correlation between population density and female territory size: $r=-0.905$, df=4, $P<0.05$

area she may have never visited. Rather than heading uphill to her home territory, she traveled farther away.

A week later, she was in a beech-hemlock forest 1.4 miles to the east, where she remained for 2 weeks. But when I visited her on May 30, 1 found the ground under her rest tree strewn with many dozens of porcupine quills, far more than would be expected from loss in grooming. Because the quills were widely scattered, they could not be explained by the porcupine's falling out of the tree. I concluded that Gro had fought with either a would-be predator or a resident porcupine. From the number of quills lost, I also concluded that the fight had been a serious one.

Apparently, Gro had been unsettled by the experience because a week later she abandoned her location in the hemlocks and moved 2 miles west, climbing more than 1000 feet in attitude. I found her on a southern ridge of Vly Mountain, feeding in a beech tree. Two weeks later still, she traveled a mile north to Jaeger Ridge, completing a large circle back to the heart of her 1983 territory. I found her resting in an ash tree, with niptwigs scattered on the ground. But again, dozens of quills were strewn under her tree. Such quill-scattering is extremely rare, perhaps because porcupines are invulnerable to most local predators. In my 25 years of radiotracking porcupines, I have found only five massive quill-scatters that did not result from tree-falls; two came from Gro, and within a 3-week period. In both cases, Gro was a new arrival, first as a strange porcupine in the hemlock forest and then as a long-absent owner in her old territory. I therefore guessed that both altercations involved other porcupines, probably the resident female in the hemlock forest and a usurping immigrant in Gro's former territory. Following her return, Gro never again left her traditional boundaries, and I found no further evidence of battle.

Porcupine territorial constraints break down in two situations: feeding in fruit trees and chewing salt at the salthouse and outbuildings. In both cases, several porcupines may visit at the same time even though their territories do not overlap at other times, perhaps because fruit trees and salt sources represent limited resources that belong to no porcupine in particular. That is to say, no porcupine feels invaded when it finds another porcupine using the resources; both are visitors. (That does not prevent some expressions of irritation in the form of squawking.) The behavior may be facilitated by the location of the orchard and the salthouse outside the forest proper.

How do these observations compare with other studies? Rick Sweitzer (2003) has studied territoriality among porcupines in the Great Basin desert of Nevada. Within the predominant habitat of desert shrub and grassland, porcupines were concentrated in 3 small grove areas with permanent springs and thick juniper-sagebrush shrubs. Both males and females maintained relatively exclusive core home range areas; males showed significantly less overlap with other males than females did with other females. In their territorial intolerance, the Great Basin males differ from Catskills males. To explain the considerable territorial overlap between females as well as between males, Sweitzer hypothesizes that sharing was forced on the animals by the limited availability of the favored grove habitat. In this respect, the Great Basin porcupines are behaving like Catskills porcupines when exploiting a resource in short supply.

Juvenile Dispersal

Juvenile dispersal is defined as a one-way movement of a sexually immature animal away from its mother's territory (natal territory). A great deal of social structure derives from the pattern of juvenile dispersal; it determines who inherits the parental territory and who leaves home. The pattern among porcupines is unusual among mammals: female offspring leave, males stay (Greenwood 1980). The evidence for the pattern is based on the following case histories.

- As already mentioned, Rebecca's 1985 son, Ram, was recaptured in 1986 inside his mother's territory. After being radiotagged, he lived inside his mother's summer territory, with range extensions to the west. Although he matured sexually in summer 1987, his range showed little change until summer 1988 when he reached adulthood and began to expand his range.
- Moth's son Pleiades, radiocollared in July 1983, separated from his mother in October but remained within her territory. In November, he entered a rock den where he either dropped his radiocollar or froze to death in January 1984. The den emitted a weakening radio signal for a year and a half afterward, but I was never able to recover the transmitter.
- Squirrel's 1988 son Quill remained in his mother's territory till June 1989, when I found him dead of unknown causes.
- Moth's 1988 son, Woolbear, likewise remained in his mother's territory through December 1988. His signal disappeared later in the winter; he may have been killed by the owner of a trailer under which he was sheltering.
- Rebecca's 1984 daughter, Becky, was radiotagged in August 1984. In September, she left her natal territory and traveled 1.5 miles west across two mountain ridges, establishing a new territory in a hemlock forest far from her mother. Her radio failed in November 1984 and further contact was lost.
- Squirrel's 1987 daughter, Eft, dispersed on September 26, in a heroic one-way trip to the east. Along the way she crossed a county road, a state highway, and a good-size stream. She then climbed Saddleback Mountain and came to rest 4.2 miles from the center of her natal territory. She spent the winter exploring a series of dens; her radio failed in March 1988.
- Bini's 1987 daughter, Chili, dispersed on September 25 and traveled 1.3 miles northwest, across the ridge of Vly Mountain, to a winter territory near that chosen by Becky 3 years previously. She survived her first winter alone and became a mother in 1988. (See Figure 10.5)

Figure 10.5. A newly dispersed juvenile is fending for herself, feeding on beech slash. (Photo by UR)

All of the case histories listed date from pre-1989. Regrettably, I have been able to add only two other case histories in the 18 years since. This is due, most importantly, to the severe reduction in number of radiocollars assigned to juveniles in my study after 1988. One of these was given to L7, Loretta's daughter of 1997. I first captured and marked her in August 1997 when she was sheltering near her mother and weighed an undersized 1.4 kg. I released her without a radiocollar. But I captured her again and collared her in May 1998 when she was seeking salt at the salthouse. Over the course of 1988, L7 ranged in a territory due south of her natal territory, on the opposite side of Roarback Creek. Her signal disappeared after October 1998. Although her natal and dispersal territories showed no overlap, the centers of the two territories were only 0.6 miles (1 km) apart, less than the dispersal distance of any other female in my series.

The second post-1988 case history involves B92, Bini's son of 1992. I never radiocollared B92, but captured him twice with his mother in 1992 and marked him with a groin tattoo. In 1993, I captured him twice at the salthouse, both times in the company of his mother! The mother and son traveling as a couple constitutes clear evidence for nondispersal.

In summary, the evidence for female-biased juvenile dispersal in my porcupines is based on the life histories of 9 animals, 8 of whom carried radiocollars.

The 5 males remained in their natal territories; the 4 females dispersed. The probability of such an outcome resulting from chance alone is 0.0079 (Fisher's exact test), well below the cutoff of .05 that biologists use to accept significance.

What prompts the juvenile females to disperse? In some mammalian species, such as the wolf (Mech 1987), mantled howler monkey (Crockett and Eisenberg 1987), and gray langur (Pusey and Packer 1987; Struhsaker and Leland 1987), dispersal follows aggression from resident pack members. That does not happen in the porcupine. As noted in the previous chapter, the juvenile, whether male or female, becomes increasingly adventurous as the summer progresses, straying ever farther from its mother at night. Finally, contact between the two is broken completely; females disperse while males remain in their natal territories. The break comes during the fall mating season, when mothers stop lactating and go into estrus. The night before Chili dispersed, I was sleeping in the woods near her feeding tree. Her mother's signal was coming in weakly, from far away. Twice during the night, Chili gave a mewing call that juveniles use to solicit suckling. Her mother did not respond. The next day, Chili began her 3-week journey to the west. Hence, the mother's influence on juvenile dispersal is, at most, indirect. The urge to disperse appears, instead, to follow an inner development signal (or ontogenetic switch, in biological jargon). Such a mechanism appears to operate in other mammalian species as well, including ground squirrels, *Spermophilus beldingi* (Holekamp 1986), and prairie dogs, *Cynomys ludovicianus* (Fagerstone et al. 1981).

Although juvenile females disperse during the breeding season, they do not generally breed during their first year. Of 14 females examined in 1988 (9 adults and 5 dispersed juveniles), all the adults were lactating but only 1 of the 5 juveniles was producing milk. The sole exception was Chili, who gave birth to a male infant around June 15. That means she conceived her baby around November 15, some 50 days after leaving her natal territory. Therefore, even if a female breeds during her first year, there is little chance she will mate with her father. Similar results were observed by Sweitzer and Berger (1998), who found only 1 of 8 juveniles becoming pregnant, while 8 of 8 yearlings and 45 of 50 adult females did so. In my study, the average dispersal distance for the three dispersing females was 3.06 km (1.9 miles), corresponding to 3.2 male territory diameters.

Adult Dispersal

All the dispersal episodes mentioned so far have involved juveniles. Only dispersal by juveniles is likely to prevent inbreeding. To complicate matters, porcupines can also disperse as adults. Whereas juvenile dispersal is predictable, female-biased, and triggered by an ontogenetic switch, adult dispersal is rarely observed, male-biased in my study, and triggered by sexual competition. Four of my adult males (out of 19 males radiocollared a minimum of 3 months each for an aggregate of 25.6 porcupine-years) made permanent long-distance movements. They left established territories and never returned.

The first male disperser was Eight, a medium-size adult captured and eartagged in July 1983. Almost a year later, in May, I radiocollared him while he was molting and looking scruffy. For 2 months, he ranged in a long narrow east-west territory. Then, in July, he disappeared over the high ridge of Vly Mountain. Although I could pick up his signal at the top of Vly Mountain for several weeks afterward, the signal disappeared when I descended into the valley. I concluded he had crossed over the next range of mountains 3.6 miles to the west. After that there was only silence at his frequency, despite long expeditions to locate him.

The second permanent relocation involved the adult male Homer. When first captured in July 1984, he held a territory that almost totally overlapped that of Finder. In December 1984, Homer walked approximately 2 miles east to a hemlock forest, where he spent the winter. When spring came, Homer stayed in the new area, expanding his territory to the south. My histories of Eight and Homer are too skimpy to guess why they left their established home ranges.

Matters are different with the third known male disperser, Finder. Born in summer 1982 and radiocollared during his first winter, he became sexually mature in 1984 and achieved dominant status by 1986. During 1986, he achieved the impressive body weight of 8.0 kg (17.6 pounds), occupied a large nonwinter territory (70.2 ha), and had good mating success (I observed him guarding three different females during the mating season). Finder's world came crashing down the following summer, in 1987, when a new and powerful male, Jasper, appeared in Finder's territory. Jasper was full-grown when I first encountered him. Even though I had never captured him before, I now began to capture him regularly, leading me to believe he was an immigrant into the area. Jasper displaced Finder from his dominant position, despite Finder's slight weight advantage. An accident may have contributed to Finder's decline. During summer 1986, he suffered a serious fall that left him temporarily unable to climb. Although his injuries healed and he resumed climbing, he may have been permanently impaired.

I did not witness the showdown between the two males, but during the breeding season of 1987 I saw its aftermath. Finder's home range showed little of the expected breeding season expansion and remained at 37 ha. Concurrently, Jasper ranged over most of Finder's former territory. Several times in the fall, I found the defeated male inside a den when other porcupines were still resting in treetops; at those times, Jasper was nearby. Most important, Finder lost his harem. He failed to guard a single female during the mating season, while Jasper visited at least three—as many as Killer, the dominant male to the south.

In spring 1988, Finder left the land of his birth and began moving east. By May, he had reached the banks of the Schoharie, the major stream in the region and an impassable barrier during the spring runoff. He moved slowly along the riverbank, staying in tree cover, moving about 1 mile per week. The rate of travel was well under his capacity; in the past, he had been able to cover that distance in a single night. By June 12, Finder had reached the outskirts of Prattsville, 4 miles linear distance from his original home range but representing a 7.1-mile travel path. I saw him there for the last time, up a white ash tree that offered both food and rest. The following week, there was silence at his radio frequency despite a 100-mile search of every hilltop and valley in the area.

The fourth adult to disperse was Rebecca's 1985 son, Ram. As noted earlier, he spent the first 3 years of his life within his natal range. He often rested in trees that his mother had used and several times sat with her in the same tree, though never during the mating season. By June 1988, Ram had attained the respectable weight of 5.05 kg, yet had never been observed guarding a female. The local male who enjoyed that privilege was Killer, whose home range was four times larger than Ram's. On July 2 the younger male began a one-way journey by traveling up the mountain, higher than he had ever climbed before. He crossed the ridge, drifted across the Little Westkill valley, then started up the next mountain on August 1. I reached him there on a hot summer day, resting in the shade of a hemlock. A nearby clearing offered a distant, hazy view of Vly Mountain, to which Ram never returned. The following week, his radio signal disappeared.

The behaviors of Finder and Ram illustrate once more the flexibility of response that is a hallmark of the porcupine. Whereas juvenile dispersal appears to be a wired-in behavior that functions to limit inbreeding and resource competition, adult dispersal is a plastic response to sexual competition. A male, defeated in battle or repressed by a powerful dominant, need not spend the rest of his life as a bachelor. He may opt to emigrate and find fulfillment in woods far distant from his birth. But the journey, full of risk, may end not in fulfillment but in death.

Dispersal Data from Other Studies

Sweitzer and Berger (1998) studied dispersal in their Nevada porcupine population. The study area was a 20-km^2 enclosed desert basin, supporting an average population of 63 ± 15.4 porcupines. Between 1988 and 1993, the authors made an annual census by capturing and externally marking or radio-collaring every porcupine in the study area and assigning each animal to one of 3 age classes: juvenile, yearling, or adult.

With essentially every animal in the population tagged, both immigration and disappearance rates could be established for each age class. Animals known to have suffered mortality by starvation or predation were not included in the "disappeared" category. The results show a striking female bias in both immigration and disappearance rates. Thus, 65% of juvenile females, but only 19% of juvenile males, disappeared. This was partially balanced by an inflow of immigrants—28% among yearling females but only 5% among yearling males. Sweitzer and Berger's demonstration of female-biased juvenile dispersal both confirms and enlarges my own observations.

Recall that my own conclusions are based on 9 life histories: 5 philopatric males and 4 dispersing females. In the Sweitzer-Berger study, the male-female difference is less absolute; not all juvenile females disperse, and a small number of males do so as well. In part, the difference may represent a larger sample size in the Nevada study, but at the same time the Nevada bar is set higher. In my own study, dispersing females traveled an average of 3.04 km to their new territories. In the Nevada study, a trip of 3 km may have simply translocated a juvenile female from one corner of the 20 km^2 study area to the other, and the animal may not have been counted as dispersed. On the other hand, the difference between the New York and Nevada juvenile male behaviors probably reflects the larger sample size (42 juvenile males) in the Nevada study.

Discussion

The picture of porcupine social structure drawn from my data is that of an animal largely solitary throughout its life yet strongly attached to a piece of land (its home range) and having well-defined relationships with other porcupines. Why are porcupines solitary? Many herbivorous species forage in herds. Folivorous herd species include the New World howler monkeys (Terborgh and Janson 1986) and the gorillas and colobine monkeys of the Old World (Struhsaker 1975; Curtin and Chivers 1978). On the other hand, other folivores such as the indri (Terborgh and Janson 1986), the sportive lemur, the koala, the phalangerids, and the three-toed sloth (Montgomery and Sunquist 1986) are either solitary or restricted to mother-offspring groups.

One of the proposed benefits of group living is defense against predators. A group has more eyes and ears to detect a predator and possibly more muscle to repel an attack. But the porcupine defense is so powerful against all but a handful of predators that an early warning is not necessary, and a single porcupine can be as effective as a herd. The strategy of self-reliance is carried so far in the porcupine that even juveniles a few weeks old can defend themselves to some extent. They have to—their mothers leave them alone during the day. Another proposed benefit is defense against aggressive conspecifics. Groups of male common chimpanzees (*Pan troglodytes*) may attack individuals or smaller groups from another clan (Wrangham 1986). There is no evidence for such predatory group behavior in the porcupine. Although male-male and presumably female-female battles may ensue, all encounters seem to be one on one. Herd living may also increase the animal's ability to discover food. Catskills porcupines become feeding specialists during the summer, when high-quality linden and aspen require diligent searching through the forest. Although those foods are rare in the forest, they do form a reliable supply. Once a porcupine has learned the location of a tree, it can with certainty navigate a return, even a year afterward. The porcupine substitutes a fine-grained knowledge of its home range for the lottery of new discovery.

At the same time, porcupines avoid all the costs associated with herd living: the requirement of increased home range, the predictable decrease in individual food intake, and the increased load of parasites and disease. It should be clear, however, that even though porcupines spend most of their time alone, they nevertheless live within a complex social network. That is particularly true of males, who recognize other males, as well as females, within their range.

The female-biased juvenile dispersal observed in the Catskills is so atypical among mammals that it deserves comment. Of the 5416 mammal species recognized by Wilson and Reeder (2005), only some 22 species show firm evidence of female dispersal (Lawson Handley and Perrin 2007). Dispersal of offspring is the norm in both the plant and animal world. The benefits of dispersal are large and widely recognized: a reduced chance of inbreeding, reduced levels of competition, and access to suitable empty habitat. But philopatry (residence in the natal territory past the age of independence) also offers important benefits to a porcupine (Waser and Jones 1983): acquaintance with relatively scarce den sites, acquaintance with scarce feeding trees and salt sources, and avoidance of predators and territorial residents during dispersal. Porcupines solve the dilemma by having it both ways. Females disperse during their first year and reap the benefits of dispersal; males remain in natal territories and reap the benefits of philopatry. Because their sisters disperse, the likelihood of inbreeding for males is also reduced. (How they avoid inbreeding with their mothers remains a mystery.) Juvenile dispersal is particularly important to long-lived animals such as the porcupine, because a female that fails to disperse risks mating with her father. While I cannot calculate the residence times of the adult males in my study, one long-lived male remained in residence for 12 years.

Clear data on adult male residence times come from the 1998 study of Sweitzer and Berger. Their population census approach provides two important reproductive parameters: age at first conception for females and average residence time for adult males. The Nevada females do not breed until their second year. And the mean residence time for adult males was 26 months, with a significant proportion staying for 3 or 4 years. This confirms that a female porcupine that failed to disperse would be at risk of mating with her own father, with all the associated genetic costs.

Sex-biased juvenile dispersal is relatively common among mammals and birds and serves to maximize the benefits of both philopatry and dispersal, but why does the porcupine diverge from the mammalian norm of male-biased dispersal? Half of the female-biased juvenile dispersers are folivores: the porcupine, the alaotran gentle lemur, the red colobus, the red howler, the mantled howler, the Thomas langur, the mountain gorilla, the Western lowland gorilla, the chimpanzee, the bonobo, and the wombat (Lawson Handley and Perrin 2007). To this list of female-dispersing folivores should be added the North American beaver (Sun et al. 2000). The resource base of a folivore is relatively thin, as discussed in Chapter 6. Nevertheless, the rare lindens and poplars sought by the porcupine have the advantage of predictability; their produce can be counted on, and their locations can be learned. Therefore, they can be defended. In the porcupine, as revealed in my study, resource defense territoriality is observed only in females. That is consistent with the fact that females depend more on their resource base than do the males. The long and draining period of pregnancy and lactation means that the female strains her resource base more than does a male. It may also be a possible explanation for the porcupine's anomalous juvenile dispersal pattern.

Several other features of the porcupine social structure are logically consistent with female-biased dispersal. In juvenile dispersal, we would expect the dispersing sex to be the one that will bear the heaviest costs in case of inbreeding. Because a female bears only a single young per year but a male may sire several progeny, a female would lose more heavily if she mated incestuously with her father. Because porcupines are long-lived and because dominant males may maintain their positions over long periods, as in the case of the great apes, such a mating might be all too probable for a nondispersing female.

That brings up the symmetrical problem of how a nondispersing son avoids mating with his mother. In 1987, the question was academic for Ram. Although his testes were descended and he was physiologically capable of fatherhood, he distanced himself from his mother while Taylor, Killer, and Mayday fought over her. Killer outweighed Ram by a factor of 1.7, suggesting that it would be years before Ram reached a social rank that allowed him to even consider incest with his mother.

It is also possible that the animals know their relatedness and avoid incest. Such inbreeding avoidance has been demonstrated in Belding's ground squirrel. Paul Sherman, who for 11 years has observed thousands of matings in that long-lived species, has never observed a brother-sister or mother-son mating (Holekamp 1986). Belding's ground squirrels, like most mammals, show predominantly male dispersal. Similar incest avoidance is shown by black-tailed prairie dogs (Hoogland 1982).

Another intriguing consideration is the sex ratio of 1.58 (117 females to 74 males) among the porcupines I have examined. A number of other sources have found a female-biased sex ratio in the porcupine (Table 10.3). That ratio nicely complements female territoriality and female-biased dispersal. A mother producing an excess of female offspring will export the more abundant sex and therefore face less competition for food or den sites. Even a son costs less than full price; its adult home range will spread over the ranges of adjacent females.

Female-biased dispersal is the norm in most human societies as well. For example, anthropological work by J. W. Wood and colleagues (1985) shows that among the Gainj- and Kalam-speaking peoples of highland Papua New Guinea, a daughter is far more likely (67%) to leave her parents' parish upon

Table 10.3. Sex ratios in porcupine populations

	Number of animals	Female/male ratio	Source
	54	1.70	Spencer 1949
	124	1.53	Spencer 1949
	124	1.14	Gensch 1946
	214	1.08	Dodge 1967
	70	1.59	Brander 1973
	50	1.38	Hale and Fuller 1996
	141	1.17	Sweitzer and Berger 1998
	159	1.06	Mabille, Descamps, et al. In press
	191	1.58	This study
Total	1127		
Weighted average		1.30	

marriage than is a son (16%). The reason is rooted in the social structure of the tribes; the New Guineans practice tropical swidden agriculture, with land owned by fathers and passed on to sons. Undoubtedly, if land were owned by mothers and passed on to daughters, it would be the sons who dispersed. But for humans as much as for porcupines, no rigid biological determinism is likely to explain all social structure and reproductive behavior. For both species, there remains the option of flexible response to compelling circumstance.

11 Parasites

On a bright October daybreak, the Catskills foliage is glowing almost fiercely, and leaves continue to glow while falling on the trails. From outside the cabin, I am astonished to see a porcupine under the oak tree directly below, its orderly quill rows a rebuke to the chaos of fall. I see an opportunity to radiotag another animal for my studies. With a live trap and my thick rubber gloves, I race down the hillside to head it off before it can climb a tree. It does not move as I run up, but turns its back and erects its quills. I place the open door of a live trap in front, prod it gently with a stick, and have it trapped in seconds. The ease of capture should make me suspicious, but I don't discover the true condition of the animal until I examine it inside the cabin. A white halo surrounds the mouth, as though it had pushed its face inside a flour bag. The possibility of scabies comes to mind, but I have not yet seen the full horror of the disease.

I anesthetize the porcupine and turn it on its back to determine the sex. The entire abdominal area, normally dark and covered with short fur, is a mass of thick white encrustations, like bracket fungi on rotten wood. Deep red fissures cut into the skin. The left heel looks elephantine, swollen with two white callosities, making it impossible for the animal to climb a tree. If it tried anyway, the stretching of the underside would open the cracks in the abdominal wall. The claws of the hind feet are abnormally long, as though from disuse. I cannot determine the sex; the genital area is totally encrusted. Only later do I discover it to be a female. I name the animal Scapo, for scabies porcupine (Figure 11.1).

I have no medication on hand to treat the disease, and in any case I suspect it is too late. I attach a radiocollar anyway, to follow the behavior of a gravely sick animal, and I keep her in the cabin overnight for observation. The disease has not affected her appetite or her ability to digest and excrete. In her cage, she reduces a large apple to a few seeds and flakes from the core, and leaves an abundance of droppings, rather small but normal-shaped.

Figure 11.1. Scapo with scabies-encrusted legs and lower abdomen. She is unable to climb trees and travels only with supreme effort. (Photo by UR)

I release my captive the following morning. She is eager to get away but can move only in an awkward, three-legged gait, holding her swollen left hind foot off the ground. Her quills are strongly erected along the back and tail. She passes several climbable trees and heads toward our woodshed about 15 feet away. There she squeezes between the foundation and an open steel drum I use to root tree seedlings. In that poor sanctuary, Scapo seems to go into shock. Her quills no longer fly up when I approach and place an apple on the ground. The only signs of life are the shallow breathing movements of her sides.

On the morning of release, the woods have lost the splendor of the previous day. A night wind has ripped great openings in the forest canopy, and gray clouds race across the sky. Soon frost will kill the remaining green forage, and the porcupine will enter her most difficult period of the year. She is so exhausted, I fully expect to find her in the same spot next week, dead. But when I return in the afternoon, the apple I left has been eaten. Scapo is going to fight for her life.

My work in the city leaves time for mountain visits only on weekends, so I next encounter Scapo a week after the release. Early on Saturday morning, she is just where I left her; she wasn't there when we arrived the previous night. From above, she looks healthy. Her sides heave rhythmically; her long quills and hair glisten with molten microcrystals of snow. Yet a subtle indication of sickness remains; there is almost no quill erection as I stand over her. As an open-air den, Scapo's space between the woodshed and the steel drum offers some real advantages: acorns lie scattered over the ground around her, and the woodshed shields her from the wind. Watching her, I feel a surge of simple happiness. Last week, I thought the animal had no chance to live. Today, the odds have shifted, and the victory is mine as well. But I am to discover at what cost her life is maintained.

The next day is mild and sunny, and when I return from an expedition at noon, Scapo has left her open-air den to forage on the ground a few feet away. Her technique is like that of no other porcupine I have seen. She keeps her malformed left hind foot off the ground and slowly, by a motion almost reptilian, half walks, half crawls with her remaining feet. To feed she pushes against the ground with her snout and picks up the food directly with her mouth. I watch for an hour and a half as she slowly crisscrosses a small area near the woodshed. When she finishes and moves back against the side of the woodshed, I walk down to investigate. The ground is littered with the splintered skins of red oak acorns. Acorns are always a major fall food of local porcupines, and this year the red oaks are producing a rich mast crop. They are also an important resource to local white-tailed deer, black bears, raccoons, eastern chipmunks, and wild turkeys. A healthy porcupine has an advantage over all those animals because it can climb the oak trees and harvest the acorns before they fall. Scapo is clearly not in a condition to climb trees and will have to depend on ground foods for sustenance. Although a good bushelful of acorns lies on the ground about her, she will have to compete for them with other, more efficient harvesters. A chipmunk works from the rock slope below the cabin, industriously hiding acorns underground. More serious competitors appear at night. Deer come with the first darkness, not bothering to run away when I stand outside the cabin. Moving just outside the cone of light, they wait for me to leave. I hear their snorting when I open the kitchen window an hour later. With competitors like that, what will happen to Scapo when the last acorns have been vacuumed off by deer? Most troublesome of all, what will happen when the snows come? Can she feed on tree bark at ground level? Can she travel in the snow? Can she survive without a proper den? I decide to intervene.

A week later, I bring a bottle of hexachlorobenzene lotion. In humans, a single application over infected skin kills all scabies mites; I hope it does the same for the porcupine. I go searching for the animal. The open-air den is deserted. It is now November, and under the oaks the riches of last week's mast crop have been depleted, with most acorns chewed to shells and slivers. A few widely scattered acorns remain intact, but Scapo is not equipped to search them out. They will be small morsels for deer. With my radio, I find Scapo in minutes, a short distance downhill at the top of Roarback Field. She rests on the ground, under a small white spruce. Her disease has worsened. The white halo around her mouth has begun to encrust, and scabies has leaped to a new, vital area; white halos now encircle both eyes. She will soon be blind. She protests my close examination with small, repeated jerks of her back and tail, though the quills show only minimal erection. To apply the lotion to her belly, I flip her on her back with a small stick. She has almost no strength left. She shields her belly with her hind feet, but I have no difficulty swabbing the belly and crusted hind foot. When I try to cover the area around her mouth, Scapo bites the sponge and tries to push it away with her front paws. I am finished in a minute. The animal gets back on her feet and painfully moves away. At night, she is in her old quarters by the woodshed. I leave two apples and a peanut butter sandwich. All are gone the next morning.

Next day, the calendar reads November 2, but the mildness and sweetness of the day say early September. My daughter Rachel discovers the porcupine in the grass between the woodshed and the garden. Scapo is rubbing her jaw on the ground. Whether she is cropping grass shoots, searching for acorns, or just rubbing an itching jaw, I cannot tell. I bring her peanut butter toast and two apples and, from three body lengths away, watch as she slowly consumes them. I watch for an hour and a half, warm in the afternoon sun, and hear a long-forgotten sound, the song of crickets, a fragment of summer returned. Milkweed is blowing in a gentle breeze. As a single seed wings past Scapo, she makes a sudden, energetic snatch, catches the seed in her mouth, and crunches out the brown center. While she feeds, I carefully move closer till I am only a body length away. She tolerates me as long as an apple remains in her mouth. When it is gone, she clacks a warning with her teeth, waits, then clacks again. She crawls about 20 feet away, just inside the forest edge, and rests her head on a small pillow of moss. She lies there the rest of the afternoon and into the night as we prepare to drive back to the city.

I return at midnight a week later and find the mountainside changed. A thick, fine-grained snow is flying, accumulating in field and forest though melting on still-warm roadways. Scapo is in a new den perhaps 200 feet away from the woodshed, an enormously thick, hollow-trunked ash tree. While snow is swirling outside, the inside of the tree remains dry and earth-warmed. Most important, the hollow has a ground-level entrance. Scapo has made the tree her own; the pungent odor of porcupine rises to the opening. I leave an apple as a housewarming present. Next afternoon, the snow continues to fall. Scapo remains inside the hollow ash, consuming the apple. No tracks lead out from the entrance; she is waiting out the snowstorm before venturing forth. The storm is intensifying, and I must cut short my visit. It is the last time I see Scapo alive.

On my next visit, deer-hunting season is in progress. A layer of snow covers the ground and displays the footprints of trespassers crossing our land.

My radio apparatus picks up Scapo's signal at close range. At first the antenna appears to point to the hollow ash tree, but at the tree the signal continues from beyond. There are no tracks, no sign of recent occupancy. But as I continue, I run into unmistakable porcupine sign, a row of seven beech trees all freshly chewed along the base. I recognize Scapo's work; she has been unable to climb and feed in the crowns of the trees. Only a small amount of bark has been removed from each tree, a patch about the size of a human palm. It is far less than amounts taken by healthy porcupines and less than needed to sustain life. Across the lumber road I find another debarked tree, a young linden. I feel a foreboding. The pattern of chewing seems demented. Instead of concentrating on a single patch of bark, the tooth marks form a narrow band all the way around the tree, without penetrating to the inner bark. I imagine Scapo pushing herself around and around the trunk, scraping the bark with her incisors. It is her death tree.

The nature of the radio signal has changed. It is booming, telling me I am extremely close to the source. Then I see fine porcupine hairs projecting above the snow. I uncover the thickly quilled back of Scapo, lying with her eyes closed. The body seems fresh, the limbs not yet locked into rigor mortis. I examine her belly and see that the terrible scabs have been responding to my lotion. There are no open sores, and much of the belly is covered with fresh, hairless skin. Only the left hind foot remains unusable. Her weight is only 3.2 kg (7.1 pounds), a full 2 pounds less than 5 weeks previously. She has starved to death.

While the nutritious herbs and acorns of October were available, Scapo's handicapped condition permitted an adequate calorie intake. The winter snows sealed her fate. Tree bark has a low nutritional value, and a porcupine must eat large amounts of it to survive. Her weakened condition did not allow that. Yet she had not spent her final weeks in resignation. With determination and ingenuity, she sought out foods that she could eat, found shelter she could enter, and succeeded far beyond reasonable expectation.

Scabies Epidemiology

Scabies appears to be a relatively new disease among porcupines. In 1933, William L. Jellison of the University of Minnesota listed all then-known parasites of the porcupine: one species of louse, one species of tick, a flea, five nematode species, four species of tapes, and a tongue worm. *Sarcoptes scabiei,* the causative organism of scabies, was not on the list. The first mention of scabies in the porcupine appears in 1958, in a paper by two scientists from the University of Maine (Payne and O'Meara 1958). In 1967, Wendell Dodge wrote that scabies was epidemic among his porcupines in western Massachusetts. They were dying in their winter dens, and some of those still alive were so severely afflicted that their abdominal walls had been perforated and intestines exposed.

No other parasite afflicts the porcupine with equal severity. Although an animal may carry hundreds of lice and thousands of intestinal nematodes, it appears alert and looks healthy. In fact, there is reason to think that some of the apparent parasites may confer a benefit on their host. But in the association between *Sarcoptes* and the porcupine, the parasite grows unchecked into

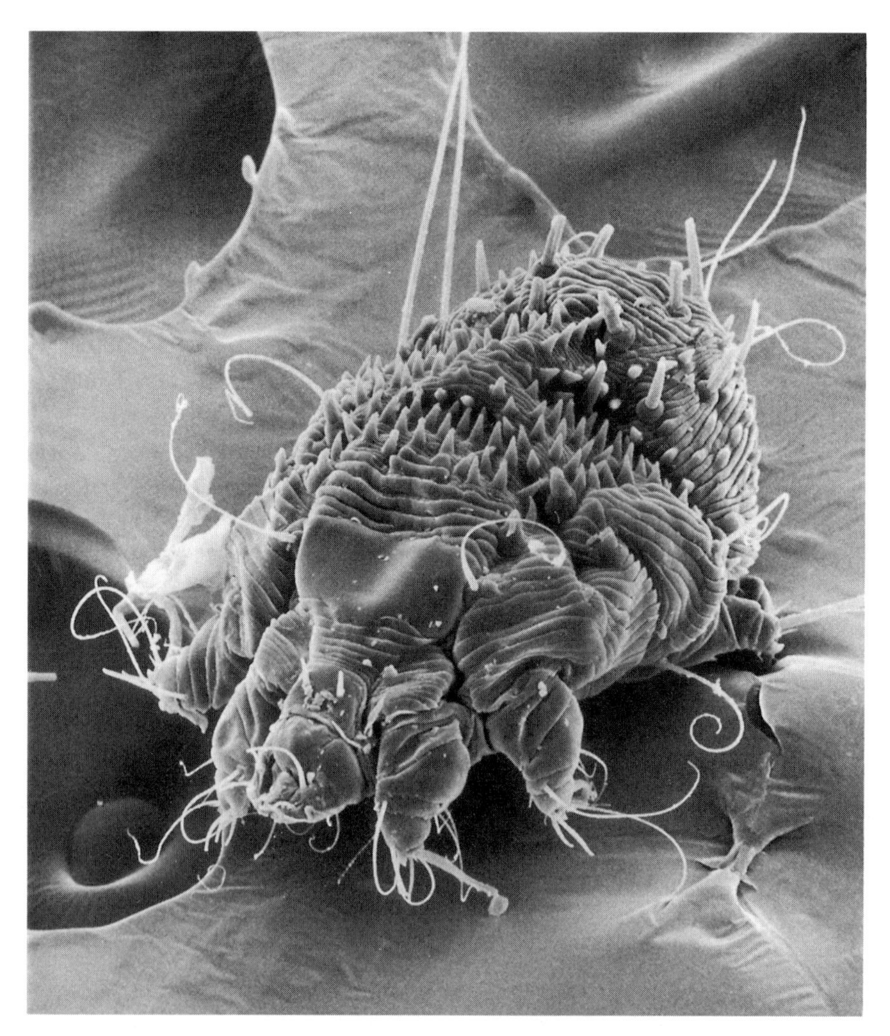

Figure 11.2. Scanning electron microscope view of *Sarcoptes scabiei,* the mite responsible for porcupine scabies. The mite burrows through the skin, setting up a hyperkeratotic response that leads to fissures in the body wall and death of the porcupine. (Photo by Enko Nonaka Yano)

the skin, causing such debilitation that the porcupine dies (Figure 11.2). One is reminded of the horrible first passage of the plague through fourteenth-century Europe or of the still-unpredictable scourge of AIDS. Perhaps in time *Sarcoptes* will evolve into a form as benign as the present nematodes and surface lice, but today the association between porcupine and mite remains in its early, lethal stage.

My experience with Scapo raised a question. If the scabies mite is spread by body-to-body contact, as textbooks of parasitology assert, how does it infect a solitary species like the porcupine? Porcupines aggregate only under clearly defined conditions: as mother and baby during the summer and, briefly, as mates during the mating season. At less predictable intervals, porcupines may share a den. In my experience, only a single male and female are involved, although under more crowded conditions several animals may share a den. Porcupines need not touch one another in a den to transmit scabies; indirect proof comes from my own intimate encounter with a denned porcupine.

On a cold February afternoon, I followed a porcupine trail into what looked like a small hole in a snow bank. I labored down the hill in my snowshoes, collected a snow shovel and flashlight, and returned to the den site to learn what an active den looks like. In the snow bank, I found a rock outcrop about 15 feet wide. Below it stretched a broad but fairly low opening, like a front porch. The den proper was in the back, with a narrower opening about a foot in cross section and extending about 3 feet into solid rock. At the end, I saw the white-tipped tail of a porcupine in the beam of my flashlight. The den was absolutely bare; the porcupine was crouching on cold rock. The porch area, where I was lying on my belly, was more cluttered. In front of the den opening lay a neat, foot-high pile of porcupine droppings, with droppings from past years on the bottom and fresh ones on the top. Scattered across the porch was a layer of other droppings, mostly old, and occasional dried leaves blown in by autumn winds.

I crawled in as far as I could to photograph the porch area and the inner den and spent perhaps a quarter hour in the same horizontal posture as the porcupine. Then I carefully shoveled snow back around the opening and exposed rock, leaving a single entrance as before. The porcupine did not find the disturbance intolerable; fresh tracks emerged from the den for several weeks afterward. Meanwhile, however, my encounter had not quite ended. A few days later, I began to itch around the chest, under the arms, in the sides. I suffered for a few days, then sought counsel from my wife, who is a doctor. She found small inflamed spots on my chest and sides and announced, "You have the scabies." Fortunately, human scabies, caused by the same organism that attacks the porcupine, is easily treated. A single application of hexachlorobenzene lotion eliminated the infection.

Contrary to textbook dogma, I had picked up the sarcoptid mites from the floor of the porcupine porch, not by physical contact between me and the porcupine. The mites had crawled toward body warmth through my thick layers of clothing. Direct transmission was impossible; I had never touched the porcupine.

My experience suggests that porcupine dens, which may be used by more than one animal in succession, may transmit *Sarcoptes* and perhaps other parasites from animal to animal. It also suggests an adaptive value to the den-cleaning activity of the porcupine. If leaves and other debris were not removed from the dens, the levels of infestation might be still higher. I tried to confirm my den-transmission theory with a standard ecological technique called a Berlese extraction. About 100 g of den droppings are loaded into a large funnel with a screen across the bottom. The bottom opens into a small collecting flask containing glycerine and 70% alcohol. The top is warmed with a 10-watt bulb. Over a week, as the debris slowly dries from the top down, the tiny arthropods migrate lower until they tumble down into the collecting flask. The technique therefore extracts only the living organisms of a litter sample. They may then be mounted on slides and examined under a microscope. I was surprised by the richness of microanimals: insect larvae, rove beetles, tiny wasps, fleas, and springtails. But the overwhelming majority of arthropods were mites (eight-legged relatives of ticks and spiders); from one den alone, I recovered 10 species of mites, with 46 individuals in total. None were *Sarcoptes,* nor did I find *Sarcoptes* in any of four other dens examined. That was not surprising in view of the small number of dens examined and the lack of assurance that any was inhabited by an infected porcupine. I could not recheck the infected den because I was running the experiment a year later.

Two years after my unfortunate encounter with porcupine-den *Sarcoptes,* another opportunity to test the den-transmission theory presented itself. I captured a scabies-infected porcupine by climbing after it into a hemlock tree; I then brought the animal back to New York for observation and treatment. I kept it in a wire mesh cage in a cool basement, collected the droppings on newspapers placed under the cage, and extracted the droppings in a Berlese funnel as before. This time, in the artificial conditions of captivity, only two mite species appeared, and one of them was the molelike, skin-burrowing *Sarcoptes:* one adult and six larvae. Clearly the mites were not passed with the feces—the organism is an ectoparasite. The mites had dropped down from the skin and fur of the infected porcupine, surviving off their host's body for the several days it took to collect the sample and make the extraction.

Afterward I cured the porcupine with a 1-minute dip in 1% malathion. The malathion solution is preferable to hexachlorobenzene because it reaches the skin between the quills and hair in a way the hexachlorobenzene lotion cannot match (Roze 2004). Most veterinarians treat scabies infections with Ivermectin, but in porcupines this drug will also remove the intestinal *Wellcomia* nematodes, which may be involved in sodium regulation in this animal.

My porcupine scabies studies have ended because the disease has run its course in my population. The cases I encountered are listed in Table 11.1. Since my last encounter with a scabious porcupine, I have made more than six hundred captures and physical examinations; not one animal has been found to suffer from scabies. The disease, at least in my Catskills population, has returned to its pre-1958 status—not observed in porcupines. But scabies epidemics are still killing porcupines elsewhere. In August 2005, I got a message from Steven Shaffer, a ranger at the Tuscarora State Forest in southern Pennsylvania, reporting encounters with a number of porcupines with advanced cases of scabies.

As noted in the previous chapter, the local porcupine population density rose significantly between 1983 and 1988 (Table 11.2). Supporting evidence that the post-1983 rise in population is real comes from the parallel contraction in female territories (see Table 10.2) and from increased intensity of linden use in 1985 (see Table 6.2). Apparently, the scabies mite, during its residency in the area, was controlling the local porcupine population. But a

Table 11.1. Scabies infections in porcupines

Year	Number of infected animals	Fate of animal
1979	1	Den encounter; animal not treated
1980	1	Hexachlorobenzene treatment; died
1981	1	(Scapo). Hexachlorobenzene treatment; died
1982	1	Advanced disease, untreated; fate unknown
1983	1	Treated by 1% malathion dip; recovered

Table 11.2. Scabies and porcupine population density

Year	Captures/ year	Days in field	Corrected captures/year	Scabies present
1983	14	116	12.1	+
1984	18	110	16.4	–
1985	26	101	25.7	–
1986	27	88	30.7	–
1987	26	85	30.6	–
1988	39	105	37.1	–

Note: Population density estimated from number of unique individuals captured per year, corrected for constant capture effort (1.0 = 100 days in the field).

parasite that is too devastating to its host is ultimately a badly adapted parasite. On a solitary species like the porcupine, the host that *Sarcoptes* kills is likely to be the last host it will have. The remaining porcupine parasites all interact more gently with their host.

The Porcupine Louse

This little brown insect, which lacks wings and eyes and travels up and down porcupine hairs somewhat like the porcupine travels up and down trees, is probably found at some time on every porcupine, often in unbelievable numbers. When I found the female Gro debilitated and near death at the end of the winter, her fur carried an excessive louse population. At times, a dozen or so of the little insects were visible on the surface, their small light-brown bodies conspicuous against the black fur of Gro's belly and neck. At home, I sprayed Gro thoroughly with 1% malathion in water, wetting her back and tail as well as her underparts. A day later, the newspaper under her cage turned brown with an accumulation of hundreds of louse corpses—I estimated at least 500 bodies. I changed the papers and found an additional hundred or so corpses 24 hours later, perhaps combed out by the porcupine. No more than a tiny fraction of that large population had been visible while the lice were actively foraging in the porcupine fur (Figure 11.3).

The porcupine louse's name, *Eutrichophilus setosus*, is aptly chosen. *Eutrichophilus* means "good lover of hair." This member of the Mallophaga (chewing lice) feeds on surface materials, not by piercing the skin. The living insect can be observed under a dissecting microscope. It will not run away if one or two porcupine hairs are introduced into its surroundings. The insect can form a loose pincers around the hair with its front pair of legs, then use its remaining sharp-clawed legs to run rapidly up or down the hair. At other times, all the legs are in motion. On reaching the end of the hair, it climbs off, mills around in a confused way, then climbs back and runs the other way. If five or six insects have only a single porcupine hair available, all will cling to the same hair and clamber over one another as they try to move up or down. The lice can also form a pincers around the hair with their mandibles, holding on with their mouthparts instead of their legs. That may be a feeding behavior; the mandibles may be used to scrape off surface grease or loose skin cells. I have placed the lice on the hairs of my own forearm and found them polite guests; they do not bite, although they are equipped to do so. All I felt was a busy wriggling on the skin, as they moved ceaselessly among the hairs. The lice seek body heat. Undoubtedly the reason so few can be seen on the typical

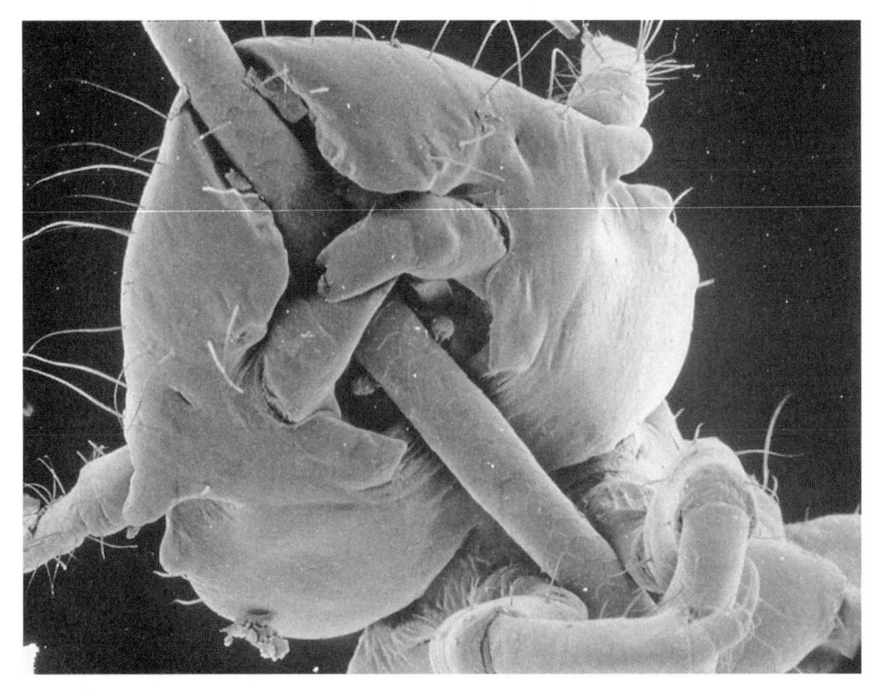

Figure 11.3. Head of a porcupine louse grasping a porcupine hair. (Photo by UR)

porcupine is because they huddle against the body for warmth. On porcupines I have found frozen in their dens, the lice have also frozen to death.

Porcupines probably acquire their lice while nursing against their mothers' bellies. All juveniles a few weeks old or older that I have examined have carried lice. The insects are more conspicuous in the shorter, darker fur of juveniles than on the adults. At that time, the eggs of *Eutrichophilus* can also be seen, glued to individual hairs like tiny white balloons. Perhaps another reason the lice are so abundant on juveniles is because the mothers molt every spring. As the adults shed their winter fur, the lice may have nowhere to go but to the fur of the juveniles, who do not molt until their second year.

It is difficult to tell whether the porcupine louse is a true parasite, because the critical studies have never been performed. I suspect that it is, in fact, a commensal (neutral effect) or a mutualist (beneficial effect) on the porcupine, rendering the service of cleaning. By scraping off excess grease and loose skin cells, the louse may prevent clumping of body fur, thus enhancing the fur's insulating function. At the same time, the lice would leave intact the protective grease layer around the quills, which are too thick for them to travel on. Such cleaning would be doubly valuable to an animal with a diminished ability to clean itself. Arguing against that view is the abundance of lice on emaciated or underweight animals. The true role of *Eutrichophilus* therefore requires further study.

Ticks

Ticks are the giant relatives of mites and the arthropod ecological equivalent of leeches. Of 70 ticks recovered from my Catskills porcupines, 67 were identified as *Ixodes cookei* by Drs. Milton Nathanson of Queens College, Durland Fish of New York Medical College, and Robert Dowler of Fordham University (Table 11.3). The remaining ticks were identified as *I. dammini* (now *I. scapularis*) (2) (Oliver et al. 1993) and *I. texanus* (1). *Ixodes cookei* is associated with medium-size mammals such as skunk, raccoon, and woodchuck (Figure 11.4). It is therefore not a specific porcupine parasite. *Ixodes texanus* is somewhat more specialized, concentrating on raccoons. *Ixodes scapularis* is known as the deer tick. Adults feed on white-tailed deer (*Odocoileus virginianus*) and often on humans; larvae feed on small mammals such as the white-footed mouse (*Peromyscus leucopus*) and on birds (Anderson and Magnarelli 1980; Carey et al. 1980; Main et al. 1982). The deer tick has also gained notoriety as the carrier of Lyme disease, caused by the spirochete *Borrelia burgdorferi.*

Acarologists in Connecticut have shown that some *I. texanus* and *I. cookei* harbor rickettsiae responsible for Rocky Mountain spotted fever (Magnarelli et al. 1985; Anderson et al. 1986). Whether the disease is transmitted to porcupines remains to be studied. Fortunately, *I. cookei* does not transmit Lyme disease (Ryder et al. 1992; Barker et al. 1993)

An adult *Ixodes*, when engorged with blood, can reach almost 1 cm in length (see Figure 2.9). When the great gray-blue body hangs from a porcupine's belly or armpit, the porcupine makes no effort to scratch or otherwise remove the parasite. Ticks do their job slowly and painlessly, but leave visible scars after dropping off.

Table 11.3. Tick species recovered from porcupines

Total	*Ixodes cookei*	*I. dammini*[a]	*I. texanus*	Identified by
33	31	2		M. Nathanson
37	36		1	R. Dowler and D. Fish

[a] Known today as *I. scapularis* (Oliver et al. 1993)

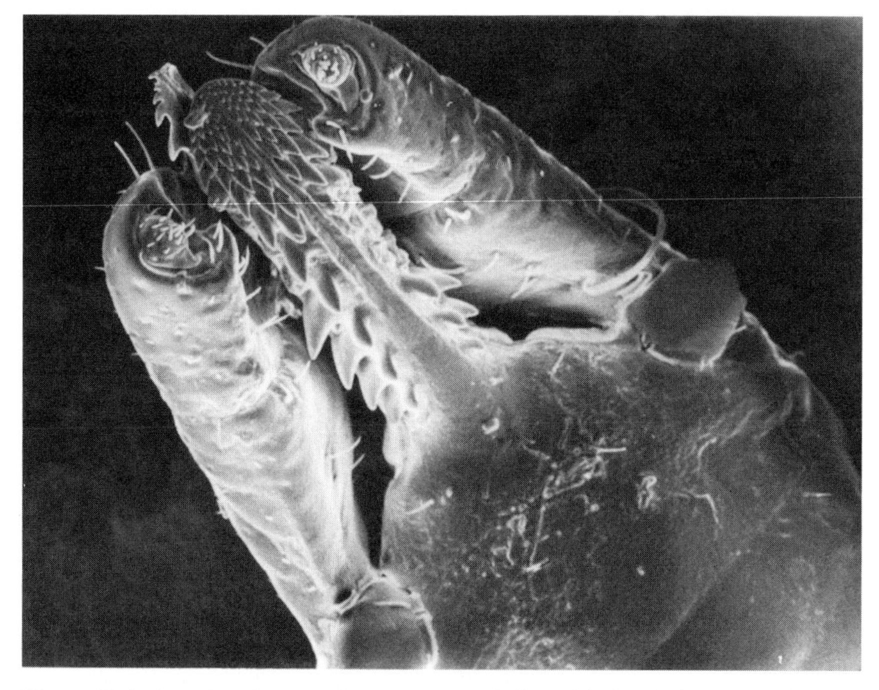

Figure 11.4. Scanning electron microscope ventral view of the head of the porcupine tick, *Ixodes cookei,* X110. The barbed hypostome anchors the tick in the skin of its host; two lateral palps serve a chemosensory function. (Photo by UR)

Catskills porcupines are essentially free of ticks from December through March. The first great wave of tick infestation (the spring wave) strikes in April and May, when up to 50% of the animals may be parasitized at any one time. Cumulative rates are higher because often a porcupine may be found free of ticks but bearing numerous scars of previous attacks. The attack rate drops to about 20% in June, then picks up again from July through September (the summer wave). A quiet period is observed during October. Then the third, smaller (autumn) wave strikes during November. The wavelike pattern suggests synchronized molting and feeding cycles, with the spring wave made up of overwintering individuals. Andrew J. Main and coworkers (1982) at the Yale University School of Medicine describe a similar multiphasic life cycle for Connecticut *I. scapularis.*

Although less than 50% of the porcupine population carries ticks even at the peak of an infestation, individuals may carry heavy parasite loads. One female carried 12 ticks in May; another carried 17 in July. They may represent hatches from a single egg mass or an aggregation phenomenon. On average, females were twice as likely to carry a tick as males (0.94 tick per female vs. 0.47 tick per male on a year-round basis). The difference is not statistically significant, but it tilts in the expected direction, since females spend more time on the ground because of nursing and maternal care. It is perhaps significant that the most highly infected male ever discovered (with 7 ticks) was Finder, who had been spending all his time on the ground while recovering from a tree-fall injury.

Ixodes ticks do not infect porcupines across their entire species range. Farther north, in Nova Scotia, the predominant tick of porcupines becomes *Dermacentor variabilis,* the dog tick (Dodds et al. 1969). In Alaska, no ticks have been recovered as porcupine parasites (Jellison and Neiland 1965).

Tapeworms

Essentially all adult porcupines carry tapeworms. Evidence comes from the white proglottids universally present in porcupine droppings and from the studies of parasitologists. Reino S. Freeman (1952) reports that 23 of 24 porcupines examined carried tapeworms. The remaining animal, although parasite-free on inspection, had been passing proglottids when first captured.

Because porcupines do not normally eat their own droppings, tapeworms must invade new porcupines through an intermediate host, an oribatid mite (Figure 11.5). Oribatids are the most common mites found in the debris of porcupine dens. They can be recognized by their convex, highly sclerotized (darkened) outer shell, which gives them the appearance of tiny beetles. As they scavenge through decomposing porcupine droppings, the mites ingest tapeworm eggs, which develop into infective cysticercoid larvae in the mites. Porcupines become infected after accidental ingestion of mites that have crawled onto green vegetation surrounding the den. In fact, the mites (14 known species) are obligate intermediate hosts for the tapeworms. Porcupines do not get infected by direct ingestion of the tapeworm eggs. Freeman collected and dissected 520 oribatids from the vicinity of porcupine dens and identified 12 infected individuals. In 929 mites collected far from porcupine dens, only 1 was infected. For good measure, Freeman also dissected and examined more than 400 porcupine lice and found no tapeworm larvae. The oribatid intermediate is common to all three anoplocephalid tapeworms found in porcupines *(Monoecocestus americanus, M. variabilis,* and *Cittotaenia pectinata).* The *Monoecocestus* tapes are almost exclusively associated with porcupines; *Cittotaenia* is rare in porcupines and better known from hares, including European species (Freeman 1949; Mead-Briggs and Page 1975).

Like many other aspects of porcupine natural history, the biological costs of the parasites have not been evaluated. On the one hand, even heavily parasitized porcupines appear to be vigorous and in good health. On the other

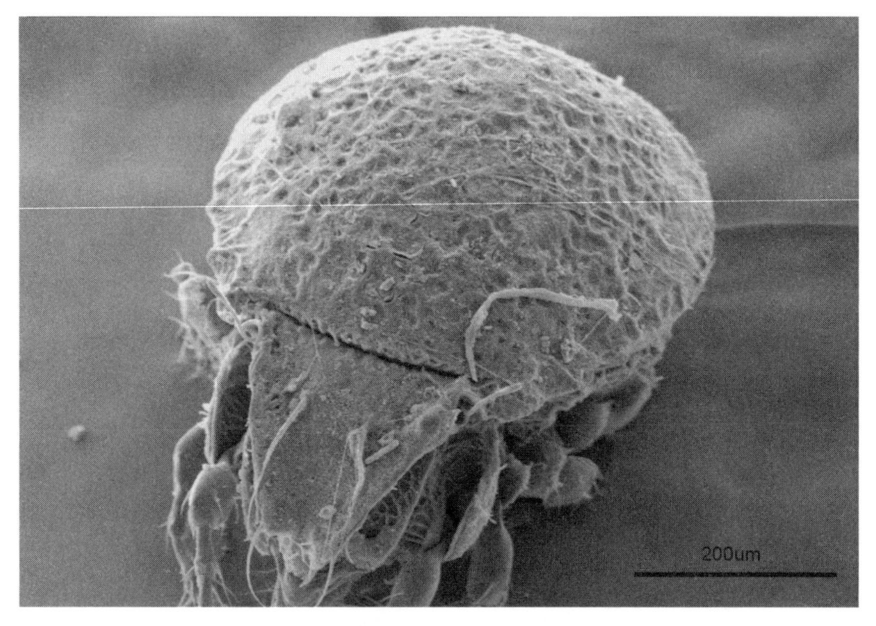

Figure 11.5. Scanning electron microscope view of an oribatid mite (*Cepheus* sp.). Oribatids, which are shielded by a tough carapace, are secondary hosts for porcupine tapeworms. (Photo courtesy of Zoë Lindo)

hand, an uncontrolled natural experiment suggests that costs may be involved. In 1947, C. A. Tryon reported on a porcupine delivered in mid-May by cesarean section and raised by hand. Its initial growth was slower than observed in wild animals, perhaps because it was not getting its mother's milk. But the growth rate picked up in midsummer; by August, the baby weighed 3 kg, and by December, its weight exceeded 5 kg. In contrast, baby porcupines in the wild first gain weight quickly and then, by late August, reach a plateau; weights achieved by the end of the first summer fall below 3 kg. Several factors may suppress growth rates after midsummer. The quality of available food is declining (as shown by declining nitrogen levels), and mother's milk becomes unavailable by October. But another factor may be the acquisition of parasitic tapeworms by early August (Figure 11.6). (Freeman showed that adult tapeworms are found in porcupines 70 days after ingestion of infected mites.) Porcupines acquire intestinal nematodes even earlier. Curtis and Kozicky (1944) report an average of 766 tapeworms per porcupine, with a high count of 1528. The tapeworms and nematodes should be absent in the hand-reared porcupine. Thus, even while wild porcupines appear healthy and vigorous, they may be sacrificing a significant fraction of potential growth to their intestinal parasites.

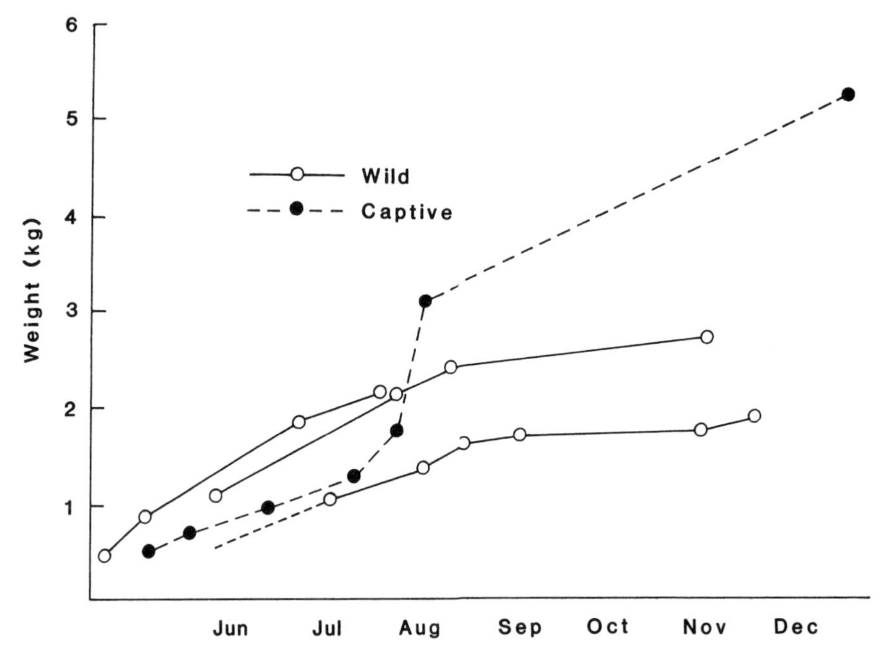

Figure 11.6. Weight gain in captive and wild juvenile porcupines. The rapid early-summer weight gain of wild porcupines plateaus around August, when they acquire their tapeworm parasites.

Nematodes

One invertebrate zoologist has stated that if all life on earth were swept away except the nematodes, the former plants, animals, rivers, and soils would still be decipherable from the characteristic massing of their erstwhile nematode inhabitants. The statement is particularly apt for the porcupine, which is known to be a host to at least four species of nematodes: *Wellcomia compar* (formerly *W. evaginata*), *Dipetalonema arbuta, Dirofilaria spinosa,* and *Molinema diacantha.*

Most nematodes are worms of microscopic size. *Wellcomia compar,* however, must be considered among the whales of the nematode world. The worms are over 2 cm (1 inch) long, relatively stout, and conspicuous because they are continually shed with the feces. Occasionally, a passed nematode may string two droppings together (Dodge 1967). In terms of body structure, the most remarkable feature of *Wellcomia* is a long lateral outpocketing (evagination) that terminates the female genital tract. The vaginas of female nematodes typically open laterally; in this case, they extend outward to give the nematodes an unwormlike shape.

The *Wellcomia* may be present in enormous numbers. Olsen and Tollman (1951) found 30,500 in a single porcupine; Curtis and Kozicky (1944) report a high count of 5184 in a single animal. The nematodes are distributed throughout the small and large intestine but not in equal densities. In one postmortem examination, I found only 6 in the lower segment of the large intestine, but there were so many thousands in the upper segment and caecum that they looked like a short-noodle soup. They became much less abundant in the small intestine and were absent from the stomach. The observed intestinal distribution of *Wellcomia* correlates with the location of the gastrointestinal sodium pool (Chapter 5), which raises the intriguing possibility that the nematodes may play some role in body sodium regulation. *Wellcomia* nematodes are also found in South American and Old World porcupines but in few other mammals (Olsen and Tollman 1951; Hugot 1982, 2002). The same pattern does not hold true with the lice, which may play a less essential role. The possibility of *Wellcomia*-porcupine mutualism deserves further investigation.

In addition to the highly visible and abundant *Wellcomia* nematodes, porcupines also carry much smaller, hairlike nematodes of the genera *Dipetalonema, Dirofilaria,* and *Molinema.* The filariid worms *Dipetalonema arbuta,* described by P. R. Highby (1943a, 1943b), may be found free in the abdominal cavity as well as in the blood, liver, and other organs. Highby shows that they are transmitted by the bite of one of several species of mosquitoes. According to C. M. Bartlett, the first four molts of the parasite are found under the skin, close to the site of injection by the mosquito (Bartlett and Anderson 1985). The fifth-stage molt migrates to the final site: the peritoneal cavity, the pericardium, or the brain of the porcupine. Other filariids are transmitted by ticks.

In comparison with humans, porcupines are hyperparasitized. A single animal may carry tens of thousands of nematodes, a thousand lice, and 400 tapeworms, plus assorted ticks, fleas, mites, and tongue worms. Some of them are unique to the porcupine, suggesting an ancient relationship between parasite and host. Both the intensity of infection and the specificity of the interaction suggest mutual benefits to parasite and porcupine. It is possible that when their relationships are fully explored, some of the parasites will be reclassified as commensals or mutualists.

12 Porcupines of the World

Porcupines other than *Erethizon* inhabit every continent of the world except Australia and Antarctica. They form two geographical and ecological groups: the New World and the Old World porcupines (Figure 12.1). The New World porcupines weigh 2–10 kg, are arboreal, and feed on tree products, including leaves, fruit, and bark. Old World porcupines may weigh up to 30 kg, are ground-dwelling, and dig for foods such as bulbs and tubers or feed on fallen fruit. They typically carry more formidable as well as more specialized quills. Both groups are classified as hystricomorph rodents, descended from a common evolutionary ancestor. (That is to say, the group is monophyletic.) Although their evolutionary relationship is not a close one, the two groups share a hystricomorph heritage: they are long-lived, have long gestation periods, and bear small numbers of precocial young.

New World Porcupines and the Origins of *Erethizon*

When the North and South American continents collided 3 million years ago, waves of animal migration swept in both directions across the newly formed isthmus of Panama. Most of the march was from north to south, the route of the deer, the horse, the raccoons, and the cats. Even the llama reached its present South American home from the north (Simpson 1980). But a number of South American animals traveled the other way. Their descendants include the opossum, the armadillo, and the porcupine.

The North American porcupine (*Erethizon dorsatum*) has traveled farther north than any of the other immigrants, yet it retains close relatives in the south—the prehensile-tailed porcupines of the genus *Coendou* and the hairy porcupines of genus *Sphiggurus*. It had been thought earlier that the North American porcupine evolved from a *Coendou*-like ancestor that migrated

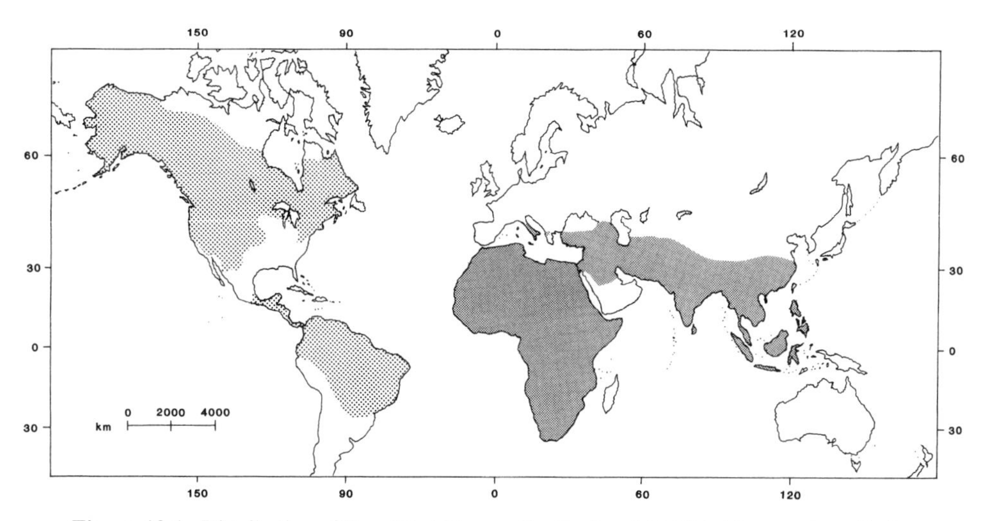

Figure 12.1. Distribution of New World (open stippling) and Old World porcupines (dark stippling). *Erethizon dorsatum* is the only porcupine of North America and has the northernmost range of all the porcupines.

north across the land bridge. But molecular dating by Vilela et al. (2009) of the split between *Erethizon* and the South American forms places the event in the late Miocene, some 7.5 million years ago. Therefore, *Erethizon* evolved in South America and migrated north once the land bridge was in place. The earliest *Erethizon* fossils in North America come from the early Pleistocene, some 1 million years ago (Woods 1973).

The New World porcupines (family Erethizontidae) include four genera in addition to *Erethizon: Chaetomys, Echinoprocta, Coendou,* and *Sphiggurus* (Table 12.1). The first two contain only a single species each. *Chaetomys subspinosus,* the thin-spined porcupine, is listed as endangered by the U.S. Department of the Interior. It was feared extinct in its Brazilian coastal forest habitat, today 95% destroyed by lumbering and agricultural clearing. *Chaetomys* had not been sighted there since 1952, but, in 1986, it was rediscovered by Ilmar B. Santos, a Brazilian biologist, who made a preliminary study of its habits (Besch 1987; Raeburn 1987). The cat-sized animal is nocturnal and lives in trees. It is covered by soft, brown spines resembling the bristles of a broom and moves with the help of a partially prehensile (grasping) tail (Figure 12.2).

A more extensive study (Oliveira 2006) has followed 3 radiotagged animals for 8–11 months in a low-canopy (Restinga) coastal forest. The average weight was 2.3 kg and average home range only 2.14 ha. The animals were invariably found in trees, resting quietly during the day and moving at night to feed, to interact with offspring, and rarely to descend to the ground to

Table 12.1. World porcupines

Old World porcupines—Family Hystricidae	
Crested porcupines	*Hystrix africaeaustralis,* Cape porcupine (lr)
	Hystrix brachyura, Malayan (Himalayan) porcupine (vu)
	Hystrix crassispinis, Thick-spined porcupine (lr)
	Hystrix cristata, Crested porcupine (lr)
	Hystrix indica, Indian crested porcupine (lr)
	Hystrix javanica, Sunda porcupine (lr)
	Hystrix pumila, Indonesian porcupine (lr)
	Hystrix sumatrae, Sumatran porcupine (lr)
Brush-tailed porcupines	*Atherurus africanus,* African brushtailed porcupine (lr)
	Atherurus macrourus, Asiatic brushtailed porcupine (lr)
Long-tailed porcupine	*Trichys fasciculata,* long-tailed porcupine (lr)
New World porcupines—Family Erethizontidae	
Thin-spined porcupine	*Chaetomys subspinosus,* Thin-spined porcupine (vu)
Stump-tailed porcupine	*Echinoprocta rufescens,* Stump-tailed porcupine (lr)
Prehensile-tailed porcupines	*Coendou bicolor,* Bicolored-spined porcupine (lr)
	Coendou nycthemera, Black dwarf porcupine (lr)
	Coendou prehensilis, Brazilian porcupine (lr)
	Coendou rothschildi. Rothschild's porcupine (lr)
Hairy porcupines	*Sphiggurus ichillus,* Streaked dwarf porcupine (?)
	Sphiggurus insidiosus, Bahia porcupine (lr)
	Sphiggurus melanurus, Black-tailed hairy dwarf porcupine (?)
	Sphiggurus mexicanus, Mexican hairy dwarf porcupine (lr)
	Sphiggurus pruinosus, Frosted hairy dwarf porcupine (?)
	Sphiggurus roosmalenorum, Roosmalen's dwarf porcupine (?)
	Sphiggurus spinosus, Paraguay hairy dwarf porcupine (lr)
	Sphiggurus vestitus, Brown hairy dwarf porcupine (vu)
	Sphiggurus villosus, Orange-spined hairy dwarf porcupine (lr)
North American porcupine	*Erethizon dorsatum,* North American porcupine (lr)

Source: Based on Wilson and Reeder (2005)
Notes: Codes in parentheses indicate IUCN status: lr, lower risk; vu, vulnerable; ?, status unknown.

defecate and urinate in specific latrine areas. Their diet was limited to tree leaves (particularly new leaves) and was highly specialized. On average, only 3 tree species accounted for 88% of each animal's diet. It appears that the distribution of one of these species, *Pera glabrata,* dictates the locations of the home ranges.

The taxonomic position of *Chaetomys* remains in dispute, with some taxonomists classifying it with the spiny rats, family Echimyidae. I here follow the 2005 taxonomy of Wilson and Reeder and include it with the Erethizontidae.

Echinoprocta rufescens, the stump-tailed porcupine, is not well known in terms of habits and biology. Its tail is hairy, nonprehensile, and about the same length as *Erethizon's. Echinoprocta* may or may not be arboreal and is found in the mountain forests of the Andes in Colombia and Ecuador.

Figure 12.2. *Chaetomys subspinosus,* the soft-spined porcupine, in a Restinga forest in southeast Brazil. (Photo by Pedro Oliveira)

Wilson and Reeder (2005) list 4 species in the genus *Coendou* (*C. bicolor, C. nycthemeris, C. prehensilis,* and *C. rothschildi*) and 9 species of *Sphiggurus* (*S. ichillus, S. insidiosus, S. melanurus, S. mexicanus, S. pruinosus, S. roosmalenorum, S. spinosus, S. vestitus,* and *S. villosus*). *Coendou* and *Sphiggurus* are so poorly differentiated that, for example, Voss and Angermann (1997) believe the two genera can not be meaningfully distinguished. But Bonvicino and colleagues (2000, 2002) maintain that *Coendou* and *Sphiggurus* do in fact have different evolutionary histories, based on chromosomal complements. *Coendou prehensilis* and *C. rothschildi* share a diploid number (2N) of 74 and a fundamental number (FN) of 82. Four species of *Sphiggurus* all share a FN of 76, but vary in diploid numbers: *S. melanurus,* 2N=72; *S. vestitus,* 2N=42; *S. villosus,* 2N=42; *S. insidiosus,* 2N=62.

Species differing in either 2N or FN cannot interbreed and hence are reproductively and evolutionarily isolated. But chromosomal complements in the Rodentia are notoriously unstable, such that closely related species may differ in both 2N and FN (Ducroz et al. 1998). Until the karyological data are joined with sequencing data, the phylogeny of the New World porcupines remains open to revision.

Coendou and *Sphiggurus* are distributed from Mexico through Central America, and in South America as far as northern Argentina. Little is known about their natural history. That is regrettable because *Coendou* is thought to be the closest living relative of the North American porcupine. A comparative study would shed light on both forms. No studies have so far been published that reveal ecological differences among the *Coendou* species or, for that matter, between *Coendou* and *Sphiggurus.*

Montgomery and Lubin (1978) of the Smithsonian Institution report on a preliminary radiotelemetry study of *Coendou prehensilis* from Venezuela. Transmitters were attached to the tail with adhesive tape and stayed in place no longer than 2 weeks. Only three animals were followed; their weights ranged from 3 to 5 kg. They were nocturnal, sleeping in trees during the day, usually 6–10 m above the ground, shielded from direct sunlight. They changed position every night, moving 200–400 m (maximal movement 950 m) per night. The home ranges were large, up to 38 ha during the 2 weeks. Male and female ranges overlapped, with male ranges being larger. The observations are surprisingly consistent with behavior expected in a similar sample of North American porcupines. Because *C. prehensilis* anatomy shows more extensive arboreal adaptations, we might have expected the animals to spend little time on the ground and to have smaller territories (Figure 12.3). One explanation for the large observed territories might be that the study site was in the llanos, a semi-open countryside where the porcupines might be forced to move longer distances in search of suitable trees. The arboreal adaptations in *Coendou* include a prehensile tail; extensions of the front and rear footpads, which, according to Pocock (1922), are movable and enormously increase the supporting area and gripping power of the foot (Figure 12.4); and muscular and skeletal adaptations for climbing, described by McEvoy (1982).

A more extensive ecological study by Estrada and Coates-Estrada (1985) on forest mammals in southern Mexico shows that the local porcupine, *C. mexicanus* (now *S. mexicanus*) has a smaller home range (10 ha), a near-total dependence on the middle levels of the tropical rain forest, and a diet consisting of approximately 80% fruits and 20% leaves. It recognizes as palatable the same leaf trees as the howler and spider monkeys. A study of *C. prehensilis* from French Guiana finds a primary dependence on the immature seeds of leguminous trees (Charles-Dominique et al. 1981). The porcupines are superb arborealists, able to reach the smallest branches and lianas and strip a tree of all its seeds. The authors speculate that *Coendou* are such efficient seed predators that the trees have evolved synchronous flowering and seed production to swamp their predators.

In the absence of detailed field studies, studies of captive animals may illuminate the biology of a species. One such study was carried out by Roberts, Brand, and Maliniak (1985) at the National Zoo in Washington, D.C. The 8-

Figure 12.3. *Coendou prehensilis* feeding by using its hind limbs and prehensile tail to cling to a branch, leaving its front paws free for foraging. (Photo by Rodrigo M. Alvarenga)

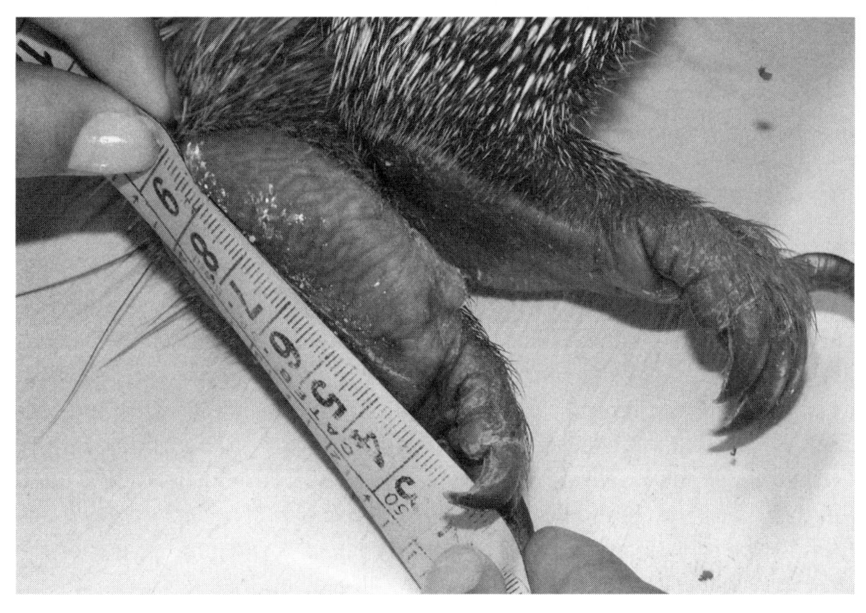

Figure 12.4. Arboreal adaptation in *Coendou prehensilis;* the large footpads of the hind foot increase the traction needed to grip tree branches. (Photo by Rodrigo M. Alvarenga)

year project followed a colony of captive *C. prehensilis* under zoo conditions. Five of the animals were wild-caught; 13 were born in the zoo. They were clearly nocturnal. Gestation was comparable to that of *Erethizon:* 195–210 days; all litters consisted of one young. At 415 g, the newborn was well developed and represented 8% of maternal weight. Infants carried a full coat of quills, had strongly prehensile tails, and had well-developed claws. They climbed readily if disturbed but generally lay quietly in one spot for the first 2 to 3 weeks of life. They nursed every 4 to 6 hours, for 1 to 3 minutes at a time; nursing provided 100% of their food for the first 4 weeks of life. During that time, they gained weight at the rate of 12 g per day. For the next 4 to 15 weeks, infants continued to nurse but also fed independently, with a weight gain of about 10 g per day. After 15 weeks, they foraged independently, reaching adult size at about 16 months and sexual maturity at 19 months. The porcupines are long-lived. One female, at 11.5 years, remains reproductively active and has produced 10 young in 8.5 years.

Perhaps the most striking *Erethizon-Coendou* difference that emerged from the study is the slow growth rate of *Coendou* infants compared with *Erethizon.* Even under zoo conditions, where the mother's diet should be optimal, it takes more than a month for the baby to double its birth weight. *Erethizon* babies double their birth weight in 2 weeks. Roberts and coworkers suggest that the difference is related to differences in the forest habitats of the two species. The temperate North American forests of *Erethizon* are highly seasonal, offering a small window of opportunity for the growth of young animals. The more constant conditions of the *Coendou*'s tropical forests allow a more gradual strategy. In addition, the observations of the Roberts group suggest that *Coendou* breeds throughout the year, in contrast to the seasonally synchronized breeding of *Erethizon.* The North American porcupine has a higher metabolic rate, which makes the rapid growth possible.

Although *Erethizon* and *Coendou* inhabit different kinds of forests and follow different ecological strategies, they are closely related as New World hystricomorph rodents and share many features of anatomy and behavior, especially in reproductive strategies. In both genera, the female has a vaginal closure membrane that dissolves during estrus and parturition. The male has a sacculus urethralis and showers the female with urine prior to copulation. There is a long gestation period (210 days in *Erethizon* and 200 days in *Coendou*), resulting in a single, well-developed young (450 g in *Erethizon;* 415 g in *Coendou*) and followed by prolonged lactation (127 days in *Erethizon;* up to 105 days in *Coendou*). Both animals are long-lived (to 21+ years in *Erethizon;* over 14 years in *Coendou*). Such similarity is not surprising in view of the close evolutionary relationship between the two genera.

It is unfortunate that more is not known concerning the natural history of *Coendou* and *Sphiggurus.* How are the genera ecologically differentiated?

How many species are there? What kinds of social systems do the animals have? (The Roberts group notes a series of complex vocalizations and scent-marking behaviors that suggest social interactions, but their natural patterns cannot be determined in a zoo setting.) How do *Coendou* interact with other leaf- and fruit-eaters, such as howler monkeys, sloths, and leaf-cutting ants? Who are their natural predators and parasites? All those questions require more study.

Old World Porcupines

Shakespeare's reference to the "fretful porpentine" (*Hamlet* Act 1, scene 5) is not to the North American porcupine but to its Old World cousin. Likewise, the porcupine emblazoned on the royal coat of arms of Louis XII of France (1462–1515) is the Old World form. So is the porcupine described by Pliny the Elder and by Aristotle before him (*Historia Animalium*). Old World porcupines have long lived in association with humans and have perhaps evolved with them. That may be why they have none of the boldness and self-confidence associated with the American animal. They are furtive, fast-moving animals specialized for ecological roles different from those of *Erethizon* and *Coendou*. Erna Mohr (1965) refers to them as the ground porcupines, distinct from the tree porcupines of the New World. Mohr recognizes five genera of Old World porcupines: *Trichys, Atherurus, Hystrix, Acanthion,* and *Thecurus*. But Wilson and Reeder (2005) list *Acanthion* and *Thecurus* as subgenera of *Hystrix;* here I follow their nomenclature (see Table 12.1). New World and Old World porcupines share a common ancestry (Huchon and Douzery 2001).

Trichys

Trichys is considered the most primitive (least specialized) of the Old World porcupines. It is small for a porcupine (1.7 to 2.2 kg adult weight), with a ratlike look enhanced by short quills hidden among the guard hairs and fur. The quills have the quality of stiff bristles, more like the quills of the spiny rats than of the higher porcupines. When frightened, the animal stamps its feet and flares out its quills. A distinctive feature is the tail, long for a porcupine, covered with scales for most of its length, and ending with a tuft of bristles. These cannot be shaken to produce a sound, as in *Atherurus* or *Hystrix*.

When the animal runs along the forest floor, it holds its tail high like a flag. Curiously, about 30% of museum specimens lack the tail entirely. There is a region at the base of the tail where it breaks easily. Various theories have been offered to explain the tail loss. One guess is that males hold the females by the tail during mating. (There are more tail-less females than tail-less males.) Another theory suggests that the natives collect *Trichys* tails for some purpose and break them off intentionally. What the special purpose is has not been declared, nor have accumulations of *Trichys* tails been discovered among the villagers. An alternate possibility is that the tail may simply be sacrificed to predators the way that some lizards sacrifice their tails to birds; the *Trichys* tail, however, cannot be regenerated once lost.

Trichys are forest dwellers and reportedly more nimble in trees than other Old World porcupines. They are nocturnal and spend the day in an underground den. They are long-lived under zoo conditions; an animal at the San Diego Zoo was still alive after 10 years of captivity (Mohr 1965).

Atherurus

The two brushtail porcupine species tend to be larger than *Trichys* but not as large as *Hystrix;* adult body weights range from 1.5 to 4 kg. They are equipped with three kinds of quills as well as an assortment of hairs, bristles, and vibrissae. The quills covering most of the back and sides are flat to hourglass-shaped in cross section and have feathery edges. The feathered (barbuled) edges are better developed on the African species, and the barbules resemble sawteeth (Figure 12.5). They clearly do not have the same function as the scales on the tips of New World porcupines, which work to pull the quill deeper into the body of the victim.

There is one function of the barbulated quills that deserves investigation—they may act as osmetrichia (disseminators of warning odor). Most osmetrichia are stiff, erectable hairs with enlarged surface area to hold the odorant. Does *Atherurus* carry a warning odor? Like all porcupines, *Atherurus* is well-defended by its quills and would benefit by advertising its defense capability.

To investigate the hypothesis, Kathleen LaMattina of the Bronx Zoo arranged a trip for me to sniff the resident *Atherurus africanus,* a 10-year-old female who is a star of the zoo's World of Darkness (Figure 12.6). The opportunity to sniff would come when the porcupine entered a small room just in back of the exhibit. When the porcupine was introduced into the room, I opened the back door and put my face in as far as possible to sniff. The calm and sweet-tempered porcupine turned, came up to me nose to nose, and sniffed *me*! Then she disappeared into the World of Darkness. I had caught no scent. But warning odor should be released only by an aroused, quill-erected porcupine. The test remained inconclusive.

Among the barbule quills, and on the African form only, emerge short, thick quills that are circular in cross section and may be lost during molting. The Asiatic form carries no circular quills. Brushtail porcupines are named

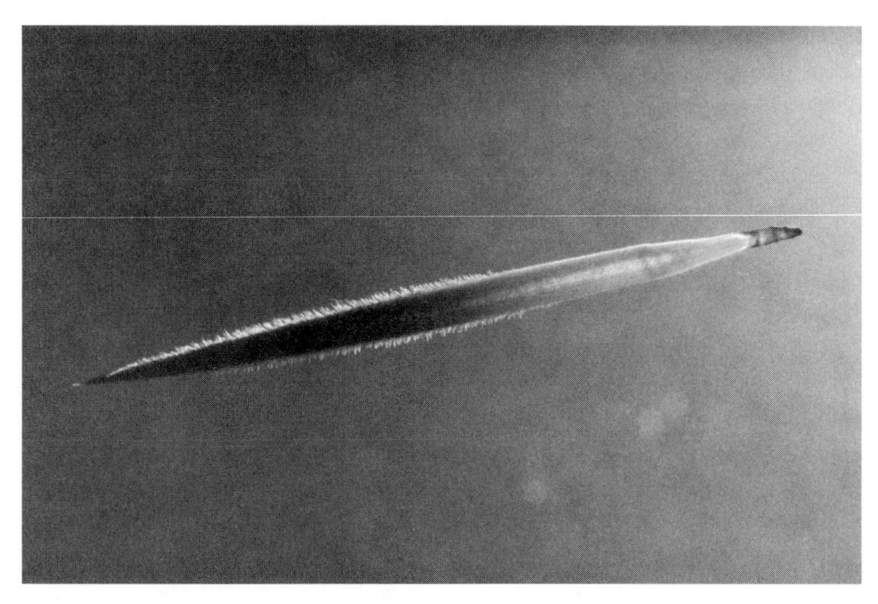

Figure 12.5. A barbulated quill from *Atherurus africanus.* (Photo by UR)

Figure 12.6. *Atherurus africanus,* holding its brush tail erect, at the Bronx Zoo. (Photo by UR)

for the large brush at the end of the tail, composed of stiff bristles and platelet quills (quills modified for rattling) (Mohr 1965). Vibration of the tail produces a dry, rattling sound that functions as a warning signal.

As in *Trichys,* the tail of *Atherurus* may break off easily. The zone of breakage occurs approximately a third of the way down the tail, where the barbule quills end and the tail carries only a few short bristles underlain by scaly rings. The incidence of tail loss in *Atherurus,* however, is much lower than it is in *Trichys.* Of 94 brushtail porcupine museum specimens I have examined, only 4 showed basal tail loss.

Reproduction in *Atherurus africanus* has been studied by Rahm (1962). The gestation period is 110 days; the single young weighs about 150 g at birth. Births are scattered throughout the year. By the second day, the newborn is able to leave its mother, go for a walk around the neighborhood, and find its way back. The mother nurses for 2 months, but babies begin taking solid food within 3 weeks.

Brushtail porcupines live in all the tropical forests of their species ranges, up to elevations of about 3000 m. They are active nocturnally. During the day, they shelter in hollow logs, rock crevices, eroded cavities in stream banks, and similar sites. They prefer not to dig their own dens, but select natural cavities that may be enlarged and modified (Mohr 1965).

Louise Emmons (1983) of the Smithsonian Institution has studied the African species by radiotelemetry. She describes the animals as nervous, fast-moving, and largely subsisting on fallen fruit. Although principally ground-dwelling, they will swim or climb a tree to escape pursuit. They are extremely wary of humans, perhaps because of strong hunting pressure. Adult males average 3.4 kg in weight and are slightly heavier than females. Babies take 2 years or longer to reach adult size. The animals are not crepuscular; they emerge from their dens only when it is completely dark. On dark nights, the activity pattern is trimodal, with two rest periods per night. On moonlit nights, all animals spend more time resting, and the activity pattern becomes bimodal, with a single rest period around midnight. They move along well-defined trails through the undergrowth.

Moonlight reduces movement for both sexes. Home areas average 13.3 ha, with no significant male-female difference. Emmons speculates that 13 ha may be the minimum area required for feeding on fallen fruit, to ensure an adequate amount and diversity of that resource.

The most tantalizing of Emmons's findings is that the home ranges of her study population fall into three separate zones of occupancy. Zone 1 is used by one male and one female, zone 2 by three males and two females, and zone 3 by one male. The usage patterns of the three zones are mutually exclusive. Emmons calls zone-sharing animals members of a clan. With Roussillon,

Emmons has collared 18 additional animals and found them all to be members of clans (L. H. Emmons, personal communication). The home range of any individual could be predicted from its capture site. Clans consisted of about 3 females and 5 to 7 males each.

Emmons studied a few captive animals. In captivity, three males shared the same den and often rested in physical contact. During encounters, animals groomed one another and made auditory displays of dominance and submission. An adult female appeared to dominate all three males. The males almost immediately killed a juvenile female placed in a connecting enclosure.

Although the African brushtailed porcupine is found alone in 89% of all sightings, it lives within strong social constraints and is well aware of other porcupines in its forest zone.

Hystrix

For 3 of the 8 species of *Hystrix (H. africaeaustralis, H. cristata,* and *H. indica*), recent studies have brightly illuminated the animals' natural histories and ecological strategies. The most extensive of these have been the studies by Rudi Van Aarde on the cape porcupine, *H. africaeaustralis.* All the *Hystrix* species share considerable physiological and ecological common ground, however, and are discussed here as a group (Figure 12.7).

Social Structure The cape porcupine, whose social structure has been most intensively studied, lives in family groups led by monogamous adult pairs. The "marriages" were stable at least for the duration of the 1-year study (Corbet and Van Aarde 1996); pair bonds appeared to be less stable in the crested porcupine, which otherwise shows a similar social system (Monetti et al. 2005). Pair members share burrows and home ranges. Because males do not compete for females, the two sexes have similar body weights of about 12 kg for adults. There is little home range overlap with neighboring family units. With direct interactions between neighboring pair groups not observed, it is hypothesized that territorial boundaries are maintained by scent marking.

Cape porcupines produce a single litter of 1–3 porcupettes each year. The litter interval is flexible to allow adjustment to climatic conditions in an unpredictable environment. The litter interval consists of a 100-day lactational anestrus, followed by 3–7 sterile estrous cycles of about 30 days each (Van Aarde 1985, 1987). The number of sterile cycles can increase or decrease depending on environmental conditions.

Sexually mature daughters do not conceive while living in their natal groups. The reproductive inhibition is not due to a failure to copulate but

Figure 12.7. *Hystrix brachyura,* the Malayan (Himalayan) porcupine. This Old World porcupine is classified as vulnerable by the International Union for Conservation of Nature (IUCN). (Copyright © Jerry Young/DK)

appears to be due to pheromonal inhibition by the dominant female (Van Aarde and Van Wyk 1991).

Reproductive Systems The sexes are hard to distinguish externally, as is also true in *Trichys* and *Atherurus*. Males and females differ little in terms of size, weight (females tend to be slightly heavier), and external markings. A true scrotum is not present in males. The penis may be palpated in its sheath and everted by pressure from behind. At rest, the opening of the penis points backward. *Hystrix* (and all other hystricomorph rodents) have a sacculus urethralis as part of the male reproductive system, with two large bristles at the bottom of the pouch. The bizarre structure, shared with the New World porcupines, so impressed Stuart Landry that he argued for a common origin for the New World and Old World hystricomorphs long before the mainstream of biologists felt ready to do so (Pocock 1922; Dathe 1937; Landry 1957). Another hystricomorph feature of the Old World porcupines is the presence of small prickles on the surface of the glans penis. Two additional *Hystrix* structures that echo structures found in New World porcupines include a baculum (shovel-shaped and about 3.4 cm long in *H. africaeaustralis*) and a copulatory plug formed by the male (freshly ejaculated semen quickly sets to form a gel)

(Mohr 1965; Tohme and Tohme 1981; Van Aarde and Skinner 1986).

The female reproductive system, likewise, follows typical hystricomorph patterns. A vaginal closure membrane disappears temporarily at estrus and at childbirth (Weir 1974; Van Aarde 1985, 1987). *Hystrix* species have relatively long gestation periods (94–112 days), produce small numbers of precocial young (1–3 young per litter with an average of 1.5, weighing 310–370 g), and have long periods of lactation. In all those patterns, the *Hystrix* species resemble the North American porcupine. Such heavy parental investment in a small number of offspring should lead to a high probability of survival for each offspring (Mohr 1965).

Digestive System All Old World porcupines have the same dental formula as the New World species: (I 1/1, C 0/0, P 1/1, M 3/3)×2=20.[1] That is, each quadrant of the adult jaw carries 1 incisor, 1 premolar, and 3 molars. *Hystrix* species show one important dental difference from New World porcupines and from the Old World genera *Atherurus* and *Trichys*—their molars are high-crowned (hypsodont) and are long-wearing. They help the animal macerate the sand-covered bulbs and tubers it digs out of the ground as part of its diet (Mohr 1965; Gutterman 1982; Alkon and Saltz 1985). The Cape porcupine, *Hystrix africaeaustralis,* is known for its habit of collecting bones (often scavenged from hyena feeding sites) and taking them back to its den for gnawing, apparently to correct calcium and phosphate deficiencies in its diet (Duthie and Skinner 1986).

The *Hystrix* digestive tract contains a prominent caecum, filled with microorganisms that help break down complex carbohydrates. The animals produce hard, dry droppings that resemble date pits (Mohr 1965; Van Jaarsveld and Knight-Eloff 1984).

Food Choice The natural diet of *Hystrix* includes subterranean foods such as bulbs, tubers, corms, and roots, as well as fruits and occasional tree bark (Van Jaarsveld and Knight-Eloff 1984). De Villers et al. (1994) studied food choice by leash-walking two porcupines. The animals did not feed randomly but stopped at specific feeding patches that were identified or protected by scent-marking. In this respect, the behavior of the Old World cape porcupine is again reminiscent of the behavior of *Erethizon.*

Quills The quills of *Hystrix* include some of the most formidable structures known in the porcupine world. I have used one pencil-thick quill of *H. indica* as a slide pointer. Quills of that size have been known to kill leopards, pythons, and other would-be predators. The stout defense quills are about 18 cm (7 inches) long. In *H. pumilis, H. sumatrae,* and *H. crassispinis,* the quills are thicker than the diameter of the front incisors. Much longer but thinner are the flexible, slightly curved outer quills. *Hystrix* are large animals; adults may weigh up to 30kg (66 pounds). When the quills are fully erected, the animal appears even larger and more formidable. The illusion of size is enhanced by the erectable bristles of the mane or crest present in *H. cristata, H. indica,* and *H. africaeaustralis.* During a confrontation, the porcupine moves rapidly sideways or backward, trying to impale its enemy.

Hystrix carry two other types of quills: flat barbule quills (see discussion in *Atherurus*) and the hollow, open-tipped rattle quills in the tail (Figure 12.8). A rattle quill begins its development as a normal, spine-tipped quill. During maturation, the tip separates cleanly and the spongy inner contents are lost; left behind is a hollow cup on a long stalk. Rattle quills, separated by stiff bristles, are found only on the tail. When the tail is vibrated, the bristles act as clappers against the rattle quills, producing a loud, threatening sound. *Hystrix* rattles its tail, stamps its feet, and flares its quills as part of its threat display. It may also chatter its teeth and produce grunting sounds (Mohr 1965; Smithers 1983). Whether to this orchestra of warning signals there is added the modality of odor production remains to be investigated.

Movements Like all other Old World porcupines, *Hystrix* are ground-dwelling and nocturnally active. Philip Alkon, who has studied radiotelemetered *H. indica* in the Negev Desert of Israel, reports that animals do not emerge from their dens when a bright moon is shining. Alkon has shown that the activity patterns are exquisitely correlated with moon phase. The porcupines are active every night, but the onset and termination of activity are timed to avoid bright moonlight. The animals den in limestone caves of a steeply sloped tributary wadi and travel 1–1.5km to reach a seasonal food

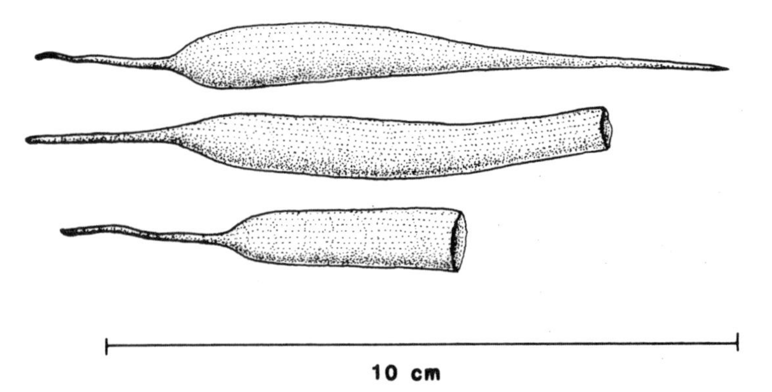

Figure 12.8. Development of rattle quills in *Hystrix*. The quills are carried at the tip of the tail and can deliver an audible threat signal.

source, the irrigated potato fields of a kibbutz (Alkon and Saltz 1988). In the case of *H. africaeaustralis,* the commuting distance from den to food source may be even longer because the animals den in rock crevices at higher elevations and their foods are usually found in the valleys. Mohr (1965) reports that *H. africaeaustralis* may spend almost the whole night traveling.

Distribution Although *Atherurus* and *Trichys* are associated with dense forests, most *Hystrix* are found in more open countryside. The genus is absent only from areas where plants are also absent, such as the Sahara and the Arabian deserts. But the animals do well in savannah and semidesert habitats, owing their drought tolerance to their nocturnal habits, their excellent water-conserving physiology, and their habit of foraging for bulbs and tubers (Van Jaarsveld and Knight-Eloff 1994). The most extensive species ranges are shown by the crested porcupines: *H. africaeaustralis* (southern Africa), *H. cristata* (northern Africa, Sicily, and southern Italy), and *H. indica* (eastern Mediterranean to India and Sri Lanka).

The disjunct European population of *H. cristata* appears to represent a recent human introduction. The case for human introduction is supported by the absence of porcupines in Italy and Sicily during Roman times, according to Pliny the Elder (A.D. 23–79). They were still absent at the time of Isidore of Seville (A.D. 560–636), who describes the porcupine of North Africa but makes no mention of its presence in Spain or Italy. At present, the *H. cristata* populations of Italy and North Africa are so similar that they do not warrant subspecies status, suggesting a recent common history. Mohr suggests possible economic incentives for the introduction of the porcupine into Italy, including the use in preindustrial times of gilt porcupine quills as hairpins. It is more probable that porcupines were kept for the table, like rabbits and guinea pigs. Porcupines are readily eaten in Italy and are often found in meat markets.

The *H. cristata* of Italy and the northern Mediterranean may not be the only porcupines introduced into Europe by human agency. Recently an unintentional natural experiment played itself out with the Himalayan porcupine, *H. brachyura.* Two Himalayan porcupines escaped from a wildlife collection in Devon, England, in 1969 and established a population in the wooded valleys of the rivers Oakment and Torridge. The porcupines damaged conifers by eating the bark, a feeding mode they also use in their native range to supplement the normal diet of bulbs and tubers (Gosling 1980). The examples of *H. cristata* and *H. brachyura* suggest that Old World porcupines retain reservoirs of biological adaptability and vigor.

Old World–New World Relationships

Are the Old World (family Hystricidae) and New World (family Erethizontidae) porcupines related, or do they represent the parallel evolution of unrelated stocks? The issue has been contended for a long time. In terms of skull characteristics, both families are hystricomorphous hystricognath rodents.[2] Hystricomorphy and hystricognathy have to do with patterns of insertion of the *masseteris* muscle of the jaw; the pattern found in hystricomorphs enhances the anterior-posterior movement of the jaw and should increase the effectiveness of the incisors and cheek teeth. The question then becomes, do the hystricomorphs have a common evolutionary ancestry?

It is fair to say that up to the mid-1970s the informed consensus favored the theory of parallel evolution. For example, such respected texts of mammalogy as those by Anderson and Jones (1967) and Vaughan (1978) supported this position.

A dramatic reversal came between 1973 and 1985. In 1973, the Zoological Society of London held a symposium on the biology of hystricomorph rodents (Rowlands and Weir 1974). In the words of Stuart Landry, "The acceptance of the term 'hystricomorph rodent' in the title . . . is a quiet signal that a long controversy is over" (1977). The research presented at the London symposium showed biological similarities between the Old World and New World hystricomorphs that no reasonable interpretation could picture as parallelism. But the 1973 symposium remained a largely indirect statement.

In 1985, the NATO Conference on Evolutionary Relationships Among Rodents tackled the question head-on (Luckett and Hartenberger 1985). The turnabout came not through new fossil evidence but through the technique of shared-character analysis.

Strong new support for the argument that the hystricomorphs of the world share a common evolutionary history comes from DNA sequence analysis. For example, Murphy and colleagues (2001) analyzed over 12,000 base pairs from 20 nuclear coding loci and found 100% support for Old World–New World hystricomorph monophyly. A similar conclusion was reached by Huchon and colleagues (2002) using evidence from 3 nuclear genes.

If we accept that conclusion, we are still faced with the problem that it is a long trip from Africa to South America, even considering that the two continents were closer during the animal transmigration tens of millions of years ago. Rafting between the two continents is not a likely event; African porcupines have failed to establish themselves on Madagascar. But other mammals of medium size have made surprising ocean journeys. Prescott (1959) observed a hare (*Lepus californicus*) travel 39–90 miles on a kelp raft. Alfred

Russel Wallace described pigs swimming up to 57 miles across ocean straits, and it is generally accepted that the New World primates arrived by a rafting mechanism from the Old World (Landry 1957). Poux and colleagues (2006) estimate the arrival of porcupine ancestors in South America as between 45 and 37 million years ago, during the middle to late Eocene.

It is not at all clear that the hystricomorph rodents that made the journey were porcupines. To say that hystricomorphs are monophyletic (have a common evolutionary origin) is not the same as saying the New World and Old World porcupines have a common porcupine ancestor. The New World chinchillas and guinea pigs are as closely related to Old World Hystricidae as are the New World porcupines. Quills have evolved several times within the hystricomorphs, not only in the Erethizontids and Hystricids but also in the Echimyids (spiny rats) and among mammals only distantly related to rodents, such as the hedgehogs and tenrecs (insectivores) and the echidnas (monotremes). Therefore, the jury is still out on the narrower question of Erethizontid-Hystricid affinities.

The porcupine families represent in microcosm the great evolutionary adventure of nature. The North American porcupine, evolved to climb trees and survive falls, to digest tree leaves and bark, and to defend itself with barbed quills, is as distinctly American as the buffalo and the wild turkey. Yet it has blood relatives on four continents who have transformed the same basic anatomy to fit physical environments and ecological niches sometimes far from those known by *Erethizon*. Despite the immediate recognizability of many of the porcupines, the animals remain fundamentally mysterious. A study of each of the species remains the best opportunity for illuminating all of them.

Notes

1. Where I=incisor, C=canine, P=premolar, and M=molar.
2. For a discussion of those terms, see, for example, Simpson (1980) or Landry (1957).

13 Travels with Musa

On a bright June morning in 2003, I embarked on a porcupine adventure that I had never experienced before: literally walking with a wild animal in the woods. Our walks lasted only 5 months, but at the end I had observed a string of porcupine behaviors that more traditional studies had given little hint of: that tree-climbing is a complex skill that juveniles must learn by hard trial and error, that porcupines routinely ingest clay (geophagy), that porcupines maintain a communications network based on a long-lasting web of urine signals, and that porcupines may ingest bits of dried feces of other porcupines (coprophagy).

My encounter begins in the Notch, a mountain pass that cuts through the central range of the Catskills. For a 4-mile stretch, the road climbs from Esopus Valley to a crest of almost 2000 feet, then plunges down to Westkill and the Schoharie. On both sides of the Notch, the mountains rise so steeply that no houses occupy the land. Because houses are absent and because the road is salted in the winter, it becomes a death trap for porcupines in the spring.

In the spring, when salt drive peaks in the porcupines, the animals surge out of the woods to the only available salt source—residual road salt around the highway. As they arrive at night to gnaw on salt-soaked twigs and plant material along the roadside, they are cut down by speeding cars. On one May 2003 weekend, I find 5 dead porcupines scattered along the Notch. Not all these deaths may have been accidental—from impact evidence, all had been hit on the shoulder of the road.

I am driving on Father's Day weekend in June 2003, following a slow-moving car. I swing around it to pass. I have traveled only a short distance when I find myself braking. In the grass on the right side of the road, a baby porcupine is moving unsteadily. I let the slow car pass me, make a U-turn, and return to the roadside porcupine. The little animal is making no effort to hide and moves in an uncoordinated way. I open the back of my car, empty my

Igloo cooler of its cold food, and then run and close the Igloo over the little porcupine. It is so weak it offers no resistance.

I reach our cabin in a few minutes and sit down to examine the little animal. When I open the Igloo, the baby is erecting its quills as far as it can. But a few minutes later, it is lying on the bottom with its legs out behind it—it seems to have spent the last of its reserves.

I slip on heavy Naugahyde gloves and pick up the baby for a closer look. Now I discover a serious injury: its upper lip has been sliced open to the gum line; the cut travels up through the right nostril (Figure 13.1). I wonder if it will be able to suckle. The injury is several days old—no blood or open tissues are visible. And the cut is relatively clean and straight, as though made by a knife. The teeth have not been chipped.

The baby is a female, and her weight is 0.85 kg (under 2 pounds), suggesting an age of approximately 1 month. Judging by her exhausted state, she has been separated from her mother for several days. Porcupine mothers do not abandon their babies—they are separated from them. In this case, I am sure the mother's body was lying somewhere along the Notch, but with state road crews regularly clearing away the evidence, I never find her body. I think back to the baby's dark shape outlined against the pitiless jumble of roadside weeds

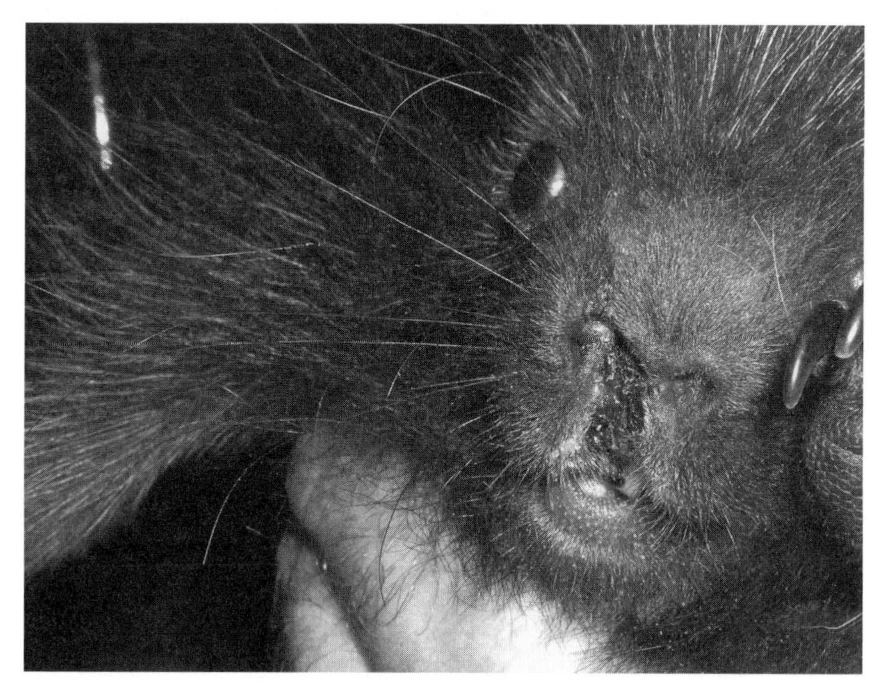

Figure 13.1. Musa with cut lip and nostril, June 2003. (Photo by UR)

and feel justified in capturing her. Without intervention, she might have become a roadkill later the same morning.

Although I have spent years studying porcupines in the wild, and have handled dozens of babies of all ages, I have never raised one by myself. Porcupines are intelligent and sensitive animals, and I realize that, if I take the baby on, I will become responsible not only for her feeding and accommodation, but for her emotional and intellectual development as well. It is Father's Day, and I decide to become a father. I name the baby Musa, after a river from my childhood. And I set to work.

Infant Care

To get some advice on porcupine baby care, I leave a message for Susan Giglia, a wildlife rehabilitator in Maine with experience in raising baby porcupines. Meanwhile, I install the baby in a cage in the basement, away from the midday heat, and drive to town to pick up supplies. Porcupine milk is thick with fat, so I buy a container of cream as an approximation. On my return, I rummage through the cabin to find a suitable nursing implement and find it—a 10-cc syringe with a plastic tip. The baby struggles as I hold her with my heavy gloves, but stops wriggling as soon as I put the syringe tip in her mouth. She sucks strongly as I press down on the plunger. The cut lip is no impediment.

Very soon afterward, I get a message back from Susan Giglia, cautioning against the use of dairy products on porcupines. Cow's milk contains a high concentration of lactose, to which most mammals are not adapted and which can cause diarrhea. She suggests diluted Pediolyte, a glucose-electrolyte mix for human infants, followed up with goat's milk, which lacks the high lactose. I return to the country store and find, to my surprise, a stock of Pediolyte on the shelf but no goat's milk.

Back at the cabin, Musa appreciates my efforts. She eagerly takes the Pediolyte from the syringe. Again, there is a struggle when I first hold her, but total relaxation when she starts to suckle. She empties four and a half syringes full of the diluted electrolyte, suggesting she was seriously dehydrated when I found her. But once she stops drinking, her struggles resume and many quills are lost. Yet with successive feedings, the baby learns to trust me to the point where she lets me pick her up with hardly a quill raised.

Feeding becomes easier when we return to New York, where the process can be a two-person operation. I hold Musa while my wife or daughter offers the syringe of goat's milk. The syringe technique is improved with a tiny rubber nipple for the tip, a gift of a Long Island rehabilitator, Barbara Bellens-Picon.

I discover in the wildlife rehabilitators a large and well-organized community across the United States and Canada. E-mail advice flows in from half a dozen rehabilitators with porcupine experience. Although there are rehabilitators who specialize in songbirds, squirrels, bats, and even bears, I find none who is a porcupine specialist. The animal is too rare and, excepting its salt supply, too independent of humans to come in frequent contact.

In New York, I install Musa in a large wire-mesh cage that lets her exercise her climbing skills. Often I surprise the little animal clinging to the side of her cage, and sometimes walking upside down along the ceiling. But at night the baby puts aside her brave face and whimpers for her lost mother. At times, her voice sounds human.

A month-old porcupine in the wild has two food supplies: its mother's milk, and the green vegetation of the summer woods. The switch from all-milk to milk-and-vegetation occurs at around 2 weeks of age. Musa's dark-green droppings indicate she has been feeding on vegetation for some time. But with the captive diet of goat's milk, her droppings become smaller and more infrequent. She needs dietary fiber, and I try to supply it with leaves of spinach, lettuce, and wild raspberry. After a lag period, the droppings return to normal size, but it is the milk she most hungers for. She lunges for the nipple when it is first presented, and suckles with little grunts of pleasure. I feel her body vibrating against my fingers as I hold her.

Change upsets her. For most of the summer, I divide my time between the city and the Catskills, and because the whole family travels together, Musa comes along in her travel cage. We make our first return trip after a week in the city. The 2.5-hour drive, with the car vibrating, lights flashing, and highway noise intruding, must be terrifying for her. We arrive in Lexington at 11 p.m., passing another road-killed porcupine in the Notch. I pick up Musa for her feeding, and she explodes. She bites my finger (fortunately, the sharp incisors don't penetrate my glove), and she pulls the nipple off the syringe. At next morning's feeding, she bites me again. Once more my heavy glove saves me, but I have learned something about the baby's state of mind.

It is a drizzly day in the country. During a lull, I take Musa to the edge of Mantis Field. I want to give the baby some exercise, and I want to test an observation by C. A. Tryon that a baby porcupine will follow the nearest moving object—generally its mother. (Tryon raised a baby porcupine delivered by C-section.) I carry Musa in her holding cage to the middle of Mantis field, then open the door, and wait. She curls up and remains still. I wait some more, then tip the cage and slide her into the tall grass. She looks to her right and left but is not tempted. She curls up even more and holds still.

I wait patiently, with no movement from Musa. Then I slowly start moving backward, sliding my feet over the tall grass to bend down the stems. After a brief hesitation, Musa raises her head and follows me. The stressful trip, the fierce attack on my glove have all been forgotten. I have been given the gift of a baby's trust.

The two of us move slowly around the field, I moving backward and glancing behind for obstacles, the baby following closely in the tall grass. Only when I reach the edge of the field, where the grass becomes thin and the dirt road begins, does Musa have second thoughts. She stops, turns, and heads back into the tall grass, and now I am following her. A raspberry leaf distracts her, and she stops for a nibble. We keep traveling until she heads toward the forest, with a woodpile at the forest edge. Its multitude of crevices and hiding places promises difficulties, and I pick her up and return home. At the evening feeding, her appetite is light, perhaps from too much excitement in the day.

On subsequent weekends in the Catskills, we try forest walks, with Musa in the lead. Her growing independence shows during a walk in July. We are taking a morning walk. It is she who sets the direction, and I who follow. Our walk has lasted about an hour when Musa decides she wants to cross a collapsed stone wall and explore the woods beyond. It is my opinion that the walk has lasted long enough, and I assume a blocking position on the other side. Musa stops for a moment and considers the situation. Then she turns around and comes at me tail-first and quills erected, the porcupine attack mode. She crosses the wall, and we continue our walk.

Fortunately, Musa's dangerous quills are seldom called into action. Although she typically turns her back when I come to pick her up, she learns to keep her quills down and never flicks her bristling tail.

In the wild, a porcupine's quills are seldom called to duty, because the animal mostly depends on its reputation for defense. But what if a porcupine lives in an area where it is the only one of its kind? This may have happened in summer 2003 during our walks in New York City, where Musa may have been the only porcupine living outside a zoo. The dogs and cats she encountered had never met one of her kind, and therefore both the porcupine and the carnivore were in danger.

I get an illustration of this on a morning in mid-July, when Musa is exploring our spacious backyard. I release her at the bottom of the yard, in a patch of grass that is overdue for mowing. She is drawn to trees, and first inspects a lilac bush with many stems. Then, attracted by the dark bulk of a Norway maple below it, she walks to the tree and carefully circles it twice. This is how she inspects the large trees in the forest.

She next works her way uphill through the tall grass, heading for a scarlet oak at the top of the garden. My daughter Rachel, who has been observing the activity, calls, "Watch out!" and points to the side. A white-and-gray cat is bearing down on the porcupine. The cat is in the final stage of a stealth attack, moving slowly with head down, tail-tip twitching, feet moving straight up and down. I have watched her use just this approach in ambushing the gray squirrels that forage under the oak tree. Yet Musa seems oblivious to the unfolding attack. I calculate that the cat would probably come out second best, with a face full of quills that I would have to explain to her owner. But the porcupine might be injured as well—she is still just a baby, with little strength or experience. I want to stop the attack before it begins, and run at the cat who holds her ground till the last second. But the slow-moving porcupine remains an inviting target, and the cat sneaks around for an attack from the other side. Her white body does not blend well with the grass and I foil her until she leaves the yard, unsatisfied.

Another high-adrenaline experience comes 2 months later, on a walk through New York's Alley Park. The park is a natural woodland, with countless trails and footpaths under the trees. Musa is scurrying along ahead of me, and I have taken my eyes off her for a moment to collect red oak acorns that litter the ground. Musa relishes them in her food bowl, but does not treat her park time as a foraging expedition. I have heard some dogs barking in the distance but am not paying attention because the park is full of dogs and dog owners. Suddenly the barking is loud and very close. I look up and start running, shouting at the top of my lungs. Musa is halted in the middle of the path, quills flared; her defensive position is seen in Color Plate 3. She is surrounded by three large, excited dogs. They are just inches away from the porcupine, and an attack could set off a fatal chain reaction: a quill in the muzzle, an enraged dog, a violent group attack. The dogs would wind up needing veterinary care; Musa could wind up dead.

My loud charge drives off two of the dogs but the third is not intimidated and stands a nose-length away, intensely excited. The owner now puffs up, and I yell at him to call the dog off. At the same time, I push the dog with my walking stick to separate him from the porcupine. This outrages the owner, who yells at me that if I touch his dog, he will do things to me that will take an operation to undo. But Musa and I hold our ground. Afterward, I think of this as a quintessential meeting between New Yorkers: loud, but careful to maintain a boundary.

The one time when Musa does use her quills, she does so in a tragic misunderstanding; her victim had meant her no harm. The scene again is Alley Park. It is mid-July, and we have just begun our visits to the park. Musa has spent an hour wandering through the woods, following no path. Perhaps she is tired—she comes to rest against the crescent base of a rotted red maple. The tree offers a shelter that only a baby porcupine would accept—her head pushed into a slight crevice but the back of her body completely exposed.

I sit down a short distance away, wondering how long the porcupine would stay, when events take an unexpected turn. The head of a chipmunk appears above the porcupine, on the edge of the trunk-opening. As I reach for my camera, the chipmunk gives an alarm whistle and dives straight down, into the tail of Musa. The tail gives a single lightning twitch, and the chipmunk disappears into a small crevice. It has been hurt—for some time I hear isolated soft whistles from its hiding place. They are the sounds of pain. I hope the chipmunk will be able to remove embedded quills in the manner of porcupines, by levering them out with its footpads.

It is now the end of July, a month and a half after Musa's arrival. Her weight has almost tripled to 2.5 kg on a diet of goat's milk and rodent chow, and she is strong enough to execute long walks in Alley Park. But I fear that in one vital area her education has been deficient—the climbing and navigation of trees. In New York state, a porcupine is above all an arboreal animal, spending most of its summer life in treetops, descending only to travel to another tree or to make a rare excursion to a salt supply. Porcupines feed in trees, rest and sleep in trees, and advertise their sexual receptiveness in trees. I have frustrated Musa's intermittent efforts to climb, afraid she might misjudge the tree and fall. As a consequence, an important geography of her young life has been blacked out.

Learning to Climb

I find a compromise; I will let her climb a tree that I can climb with her. The candidate tree is a multitrunked wild apple growing in Roarback field behind our Catskills cabin. In past years, porcupine mothers have brought their babies here to feed. Despite an abundant apple crop, none have done so this year.

The tree has thick, near-horizontal branches, and is low enough that every branch is in my reach. On a sunny afternoon, I set Musa on a low branch. A half minute later, she has fallen the 3 or 4 feet to the ground. I put her back immediately. This time she climbs, traveling a short distance vertically, then heading out on a limb that gets thinner and thinner, until she is holding on to pencil-thin branchlets at the end. Musa holds on desperately, then decides she wants to come down. But rather than descending trail-first like a porcupine, she attempts head-first descent like a squirrel. Lacking a squirrel's reversible ankle joint, she fails and returns to her tiny branches. Things get worse. Having failed in head-first descent, she tries to transfer to the branch below her. She stretches to snag it with her front limbs, but cannot reach. Suddenly she is upside down, holding on with hind feet and tail (Figure 13.2). She is going to fall, and I scramble to get below her for the catch. But a miracle takes place—she catches the branch below with her claws and transfers to it. She is now in a much better position, in the outermost branches of the lowest limb. Even if she falls, now she is so close to the ground there would be a soft landing.

Figure 13.2. Musa in a near-fall in apple tree, holding on with her hind legs. (Photo by UR)

For an animal who is supposed to be as much at home in a tree as a squirrel, Musa has given a shocking performance. And she has not even tried to feed on the apples that crowd the branches, which she gourmandizes when placed in a bowl in her cage.

The apple-tree fiasco has one concrete result—it convinces me of the importance of further tree-climbing lessons. We carry these out over the next several weeks in a series of small to medium-size trees both in New York City and the Catskills. Musa climbs a linden, a Japanese maple, and a hackberry, but our most extensive climbing is done in a plum tree in the backyard in New York. In mid-summer, the tree is crowded with plums on the verge of ripeness and is a magnet for raccoons and gray squirrels. But Musa pays no heed—she sees the tree as a three-dimensional maze to be searched. Over a succession of early mornings, she climbs, descends, and learns to turn around with increasing sophistication.

There are the inevitable mishaps. Venturing out too far on a branch, she occasionally loses her grip and is left hanging upside down, quills erect and fear-smell broadcast for all to know, until with effort she rights herself again. She does not understand the principle of path-pruning; when an experienced porcupine finds its passage obstructed by a small blocking branch, the animal nips off the obstruction and clears a path. Musa tries instead to wriggle around. In one case, she takes a nibble out of the base of an obstructing shoot, then backs off without carrying through. She returns twice to the same barrier and takes further nibbles out of the base, but does not persist to a successful outcome. In the end, the shoot remains in place and Musa looks elsewhere, turning around in seeming conflict.

On a typical morning, Musa explores each branch from base to outer tip, test-nibbles a few leaves, occasionally samples a resin exudate on the bark, then retreats to a shaded branch platform for a nap. Elapsed time from start to siesta is about 1 hour. When further movement has stopped, I gather the little animal and return her to her large, shaded cage.

Over the weeks, the tree-climbing lessons accomplish their goal. Musa no longer falls out of her tree, no longer hangs upside down from a branch, and no longer tries head-first descents down a sloping branch. Her tail becomes a sophisticated appendage for climbing support and tactile information—during a descent it flicks up and down continually to explore the environment below her.

With her mastery, Musa loses interest in the plum tree, the kindergarten of her tree-climbing. As soon as I place her on a branch, she moves to the trunk and descends, tail tapping. I respect her wishes, and we go walking in the woods instead. As often as possible, we do this in real woods, in the Catskills.

It is now mid-August, the season of ripening Catskills fruit. The first wild blackberries are turning dark, and in our garden blueberries crowd the bushes, tomatoes are half-red, and peas have matured past the soft-shell stage. There is a softness and ripeness everywhere. I pluck ripe apples from our early-apple tree while a whitetail deer watches from five body lengths away and does not scare. Musa has learned to cope with the long car trip and arrives calm (Figure 13.3). I install her in enlarged quarters—a converted woodshed with an open picture window formed of wire mesh. At night, she can watch the moon and the brilliant Mars of summer 2003.

She also gets an introduction to the local wildlife. Raccoons visit at night, attracted by her bowl of rodent chow, but the wire mesh keeps them out. But chipmunks, who come in the mornings, squeeze through the mesh and shamelessly steal her food; Musa seems unconcerned. Fortunately for her, a family of black bears that forages for ants and grubs under nearby rocks, stays respectfully away.

Geophagy

In this world of fruit and their multiple use by wildlife, Musa's walks develop a different pattern. Despite berries and mast available for browsing, she does not behave like the other wildlife around her; she does not treat her walks

Figure 13.3. Musa in August 2003, with healed face but a permanently asymmetric right nostril. (Photo by UR)

as foraging opportunities. The one exception is her treatment of clay—she eats it!

Forest soil is seldom accessible directly—it is most often covered by leaf litter and humus, by loose rock, or by living or dead vegetation. Musa draws my attention to the infrequent patches of open clay soil by stopping to lick or take small nibbles. She does this every time she passes the clay sites, and may plan her route to include them.

There is only a thin literature describing this porcupine behavior. In 1926, Olaus Murie described Alaska porcupines licking and ingesting mud along a riverbank. And in 1935, Walter P. Taylor dissected an Arizona porcupine and found its stomach full of mud. But for the next 70 years, soil ingestion has been described in no other porcupine population and consequently has not been studied in depth. Given this scarcity of commentary, I begin to wonder whether the Murie-Taylor observations and my own experience with Musa might be anomalies. Two events change my mind. The first is a chance discovery of another porcupine who behaves just as Musa does. The second is a rich literature on soil-eating in other herbivores.

The other porcupine is a male named Marvin who lives in California. I learn about him via an e-mail from Ann Bryant, a wildlife rehabilitator who is raising Marvin at home. Marvin had suffered a leg injury as an infant, and though the leg is healed, he cannot climb a tree and therefore cannot be released into the wild. In good weather, Ann and her porcupine go on long walks in the woods. I send a query: Does the porcupine eat clay?

Ann replies, "Each and every time we go on a walk, he eats dirt. He will either pick up a clod and slowly devour it, or will chew on a clear patch by a tree. He does not like it unless it's clean (without old pine needles mixed in). He will be content with this for as long as half an hour. . . ." Ann goes on to say that Marvin also likes to eat clay from streambanks. She estimates that on a single walk, he might eat between one-half to two-thirds of a cup of dirt.

The reason for the near-silence of the literature now seems obvious. There have been very few trained biologists who have followed a porcupine walking through the woods. Yet there is a rich literature on geophagy (soil-ingestion) in animals other than porcupines.

Animal geophagy has been described in birds (especially parrots and macaws), elephants, livestock, deer, giraffe, rabbits, and primates (including howler monkeys, macaques, chimpanzees, gorillas, and orangutans). And porcupines!

A variety of theories have been offered to explain the basis of the behavior in animals, and perhaps different explanations apply to different animals. Thus, elephants almost certainly ingest soil as a sodium source. Joseph P. Dudley and coworkers who studied geophagy in the African elephant in a Kalahari-sand habitat in Zimbabwe, showed that soils consumed by elephants differed from other soils primarily in their high sodium content (see Holdo et al., 2002). They also showed that intensity of geophagy was inversely correlated with body sodium levels as reflected in fecal sodium. And finally, they showed that elephants living in habitats enriched in sodium did not eat soil.

A similar conclusion was reached by David Brightsmith and coworkers in Peru, where more than 20 species of birds visit riverbank sites to ingest clay. The authors found that the only consistent distinction of clays selected by the birds was a higher sodium content compared to the clays not used.

A second, competing theory holds that clay ingestion by herbivores serves to reduce the toxicity of the animals' plant diets. A generalized plant diet imposes severe problems on its herbivore consumers. First, plants are typically deficient in sodium, an essential nutrient. This is why almost all herbivores must engage in sodium (salt)-seeking behavior. Second, plants do not stand around waiting to be eaten—they invest more or less heavily in chemical defense systems that penalize or kill would-be herbivores. The defenses may include metabolic poisons (a very large class), digestion inhibitors, hormone disrupters, photosensitizers, hallucinogens, neurotoxins, and a fiendish list of miscellania. Specialized herbivores such as many insects can counter the primary plant defense with an evolutionary arms race—they can develop metabolic pathways to neutralize the plant toxin.

But herbivorous mammals, which typically feed on a broad menu of plant species, cannot depend on this option. Their metabolic defenses must be broad-based, effective against a range of plant toxins. For example, most ungulates have a rumen, housing a broth of anaerobic bacteria and protozoa. These not only do the heavy lifting of digestion, but help break down many plant anti-herbivore products. Yet the fact that cattle are regularly sickened or killed after eating poisonous plants shows that the rumen detoxification system is not foolproof.

Other flexible detoxification strategies used by herbivores include tannin-binding proteins in the saliva, hydroxylation and conjugation enzymes in the liver, and geophagy. Geophagy is a strategy fundamentally different from the others; the animal does not produce a metabolic product to neutralize a toxin, nor does it metabolically alter the toxin to make it innocuous.

Ingested clay by itself can neutralize a major class of plant toxins, the tannins. Tannins complex with plant proteins and make them unavailable to the herbivore. Plant tannins are normally sequestered inside a plant cell, but bind surrounding proteins when the cell is disrupted by herbivore attack. Plant foods, especially in the winter, already contain such dangerously low levels of protein that all porcupines lose weight at this time. Any herbivore feeding on a tannin-laced plant risks reducing its protein intake further. That is why so many antiplant defenses focus on antitannin countermeasures, of which the salivary proline-rich proteins form one example. Although the salivary proteins are typically produced only when needed, the process is expensive—the animal must expend its own proteins in order to assimilate other proteins.

Clays represent another solution. Like the fairy-tale pumpkin that served in lieu of a coach, clays can do what the proline-rich proteins can do—they can complex dietary tannins. Geophagy can therefore become a low-cost alternative to the synthesis of salivary proteins. For example, Johns and Duquette reported in 1991 that clays reduce the tannic acid content of acorn meal by 77%. Clays can also absorb plant alkaloids over a wide range of physiological conditions. Hence, South American consumers of wild potatoes routinely eat clay with their potato meals. The result is the clay absorption of the toxic glycoalkaloid tomatine.

Clays have not been overlooked by the health-care industry. Kaopectate, a popular antidiarrheal medication, contains kaolin as its active agent. Kaolin is an aluminum silicate clay, widely distributed in nature. It binds both the diarrheal bacteria and their toxins, and the therapeutic dose is surprisingly small—less than a gram of clay per pill.

How important is geophagy to porcupines in general? Much observation and testing remain to be done, but a preliminary test in the Catskills shows Musa is not an exception. I reasoned that any clay ingested by porcupines should stay in the digestive tract and be excreted with the droppings. In 2004 and 2005, I collected clean droppings from 13 porcupines. I burned these at 750°C to remove all organic matter, then washed them with dilute acid to remove the soluble salts of calcium, potassium, and other cations derived from plant foods. The residue represented ingested clay. All porcupines had clay in their droppings, with an average concentration of 0.9%. The seasonal pattern showed lowest clay use in winter, with concentrations rising through summer and peaking in the fall (Figure 13.4). This roughly parallels the change in plant defensive chemistry through the year.

But what if the clay is being used as a sodium source? To test the sodium-supplementation hypothesis, I independently measured the time course of salt drive in my porcupines in 2004 and 2005. This was done by measuring salt consumption at my salthouse, as described in Chapter 5. As shown in Figure 5.3, p. 72, the patterns of salt consumption were similar in these 2 years, and their annual time course differs significantly from the time course of clay ingestion. Pending more direct proofs of a detoxification function for clay in porcupines such as Gilardi has provided for parrots (Gilardi 1999), the present data favor the detox hypothesis simply because it is inconsistent with the sodium-supplementation hypothesis.

Thus, geophagy may affect not only the kind of plants that porcupines may eat, but may also define the kind of territories they need to survive. This behavior, overlooked for over 65 years by standard mammalogy techniques, was

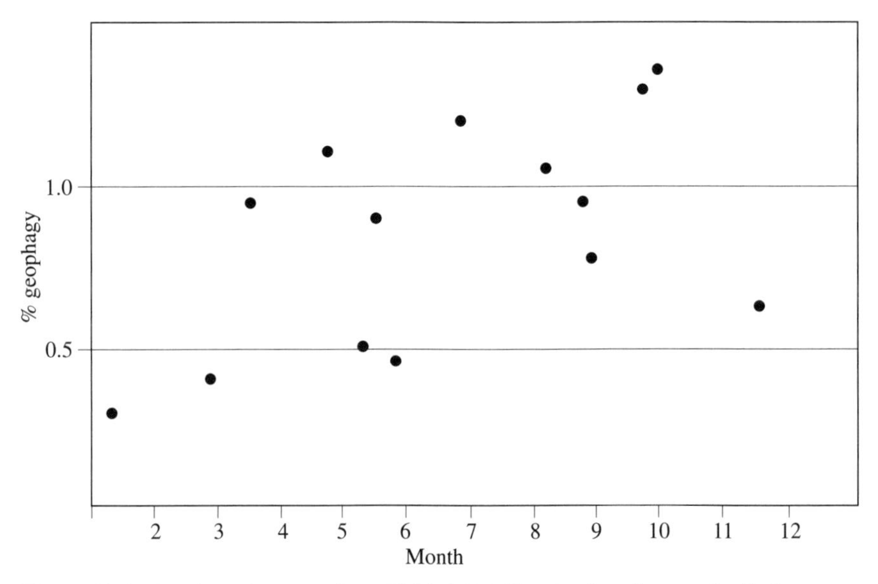

Figure 13.4. Geophagy in porcupines, 2004–2005. Clay feeding rises gradually from winter through fall and shows great individual variability. Compare this with 2004–2005 salt use in the same population (Figure 5.3, p. 72).

brought to light by following a porcupine in the woods. But she has more surprises in store.

Normal Flora

By end of August, Musa's weight has reached a respectable 3kg, equal to or better than the weights of juveniles raised in the wild. I consider the pros and cons of releasing her now. On the plus side, she would have a month to locate a winter den before cold weather forced her to do so. And release now would turn her free in a woods still filled with high-quality food: a rich acorn crop in the oak trees; green leaves in the linden and aspens; and abundant blackberry and raspberry stands, sought by porcupines for their leaves.

There is also a downside. Musa would be on her own about 2 weeks earlier than her wild conspecifics, who use the time to suck the last of their mothers' milk and perhaps pick up some wisdom of the woods. I decide the positives outweigh the negatives. But I take out insurance; I outfit Musa with a radiocollar that will allow me to track her and make sure she is all right (Figure 13.5).

I radiocollar and release Musa in the Catskills at the beginning of the long Labor Day weekend. She is angry with me because of the manhandling required to attach the radiocollar. She does not want the appendage and shakes herself from side to side to dislodge the collar, but it stays on her neck like a

Figure 13.5. Musa with radiocollar, prior to first release. (Photo by UR)

blue necklace. Musa shows her annoyance with flared quills and angry flicks of the tail—normally she lets me pick her up with hardly a rippling of quills.

I release her at the spot we usually start our walks from and follow her in the evening light until deep gloom blurs her outline. She remains unhappy with her radiocollar, periodically stopping for a vigorous shake or scratching at the nylon band with her hind claws. She has been traveling in a more uphill trajectory than usual, exploring an old treefall along the way. When I abandon her to darkness, she has stopped to rest at a spot not much further than she would have reached on a regular walk.

I listen to her radio signal at midnight. It comes from the direction I last saw her and is extremely faint, suggesting she has moved around the ridge of the mountain. The signal is in the active mode, indicating that she is still traveling. And she has experienced a brief but very violent rainstorm, which dumps a quarter inch of rain in a few minutes.

The following morning, I track Musa physically, my directional antenna pointing the way. As expected, she has moved around the ridge of the mountain, not far beyond last night's separation point. I feel an unease. She is not up a tree, where a porcupine is safe and can find food, but sits hunched on the ground next to a large sugar maple. She is no longer angry at me—she lets me approach to touching distance with hardly any quill erection, and turns her head to look at me when I make a grunting call. I take a photo and say goodbye.

I locate her again on the following day and find her still on the ground, only some 100 m northwest of her previous position. She has been feeding on abundant raspberry leaves. Everything about her behavior is ringing alarm bells: the sluggish movement, the absence of tree-climbing, and the distance from good porcupine territory. But I leave her for one more day of the long weekend.

On Labor Day Monday, an all-day drizzle announces the end of the holiday. When I track Musa, I find she has broken her lethargy and has traveled a good distance up the mountain. Were she to continue, she would soon reach an area with abundant den sites and an oak-mast food supply. But against this positive development, I note that she is still on the ground, sheltering from the rain under a fallen log. A thick growth of nettles surrounds her, with no evidence of feeding.

I decide on one more physical exam before leaving her for the week. She puts up a spirited defense against capture, a good sign. But the exam reveals disquieting facts. First, she has lost 10% of her body weight in just 2 days. Second, she shows signs of incipient diarrhea. Instead of producing the dry, crescent-shaped droppings of a normal porcupine, she produces three shapeless, soft masses during the hour I have her under observation. Diarrhea in wild animals can be quickly fatal, and this finding outweighs all the favorable signals. I must take Musa back to the city and try to treat her.

I have caught her just in time. After installing her in a cage for the trip back to the city, diarrhea sets in with full force (Figure 13.6). There is an explosive liquid discharge, and by next day she has lost one-quarter of her prerelease body weight. Porcupines can and do lose this much weight due to winter starvation, a normal and manageable process. The major component of winter weight loss is fat, accumulated during the food surpluses of summer and fall. By contrast, the weight loss of diarrhea results from massive losses of electrolytes (sodium and potassium salts), followed secondarily by water. When sufficient electrolytes have been lost, nerve and muscle function are compromised. And with accompanying water loss, kidney function shuts down and uremic poisoning sets in.

Musa has reached this point on our return to the city. She has produced no urine for a whole day. And she has become unresponsive—I can touch her without a whisper of quill reaction.

I fight these symptoms by the same techniques one would use to fight cholera in a human—I try to replace the lost fluids and electrolytes by offering her 50% Pediolyte solution. Musa has enough life force left to drink the electrolyte mix, and it saves her life. Over 3 days, the diarrhea stops and kidney function resumes. Although her weight does not show a further decline, it

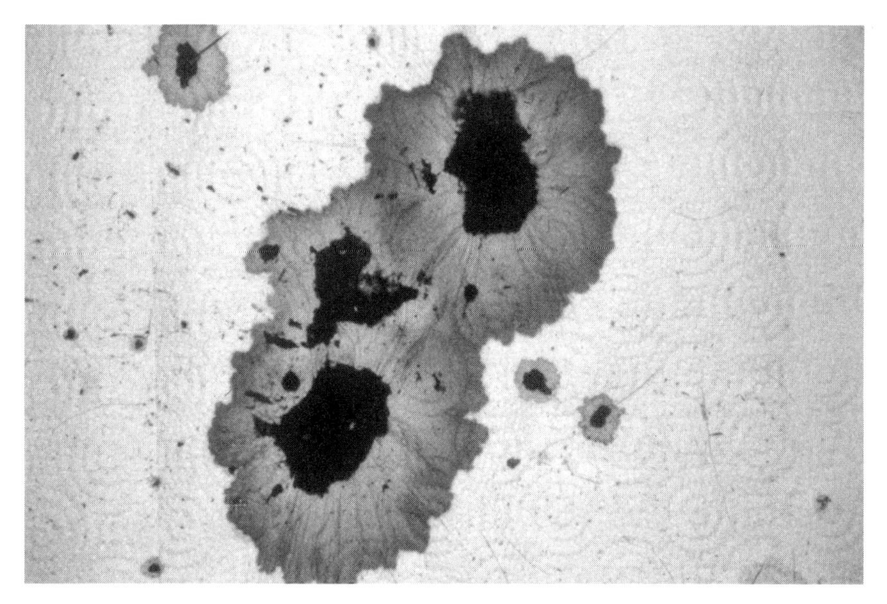

Figure 13.6. Musa diarrhea. (Photo by UR)

takes 20 days for it to start rising again, and over a month to reach her prerelease weight.

Why did Musa suffer her diarrhea attack? Diarrhea is typically caused by enteropathic bacteria such as species of *Shigella, Salmonella,* and enteropathic *Escherichia coli.* These are broadcast by the feces of wild animals. The historically important bacterium in humans is *Vibrio cholera,* spread through water supplies by cholera victims. Cholera spread in devastating epidemics across nineteenth-century America and remains a problem today in parts of Asia, but its aquatic habitat gives a safe pass to woodland animals.

All these bacteria induce diarrhea by a common mechanism—they produce a toxin that binds to a specific receptor on the intestinal epithelium. The binding sends a signal that causes the intestinal cells to undergo massive loss of sodium, potassium, and chloride ions, with secondary loss of water and intestinal contents that we call diarrhea.

In this scenario, what is the biological advantage of induced diarrhea to the enteropathic bacterium? The advantage is that the enteropathic bacteria are spread with the diarrhea. In addition, as the bowels are evacuated, so are the resident bacteria of the intestines, the so-called normal flora. The normal flora are tough competitors and do not typically let the enteropathogens become established. Then why did Musa's normal flora fail her and allow the enteropathogens their entry? The answer must be speculative, but my argument runs as follows.

1. The normal flora of herbivores are not only species-specific; they are also diet-specific. Thus, a porcupine feeding on linden leaves would have a different normal flora from one feeding on aspen leaves. The reason is that herbivores such as the porcupine depend on their normal flora for part of their digestive function; different microorganisms have different metabolic capabilities.
2. By implication, when a porcupine changes its diet, it also changes its normal flora. For Musa, the change would have been a major one because in captivity she was feeding almost entirely on rodent chow, rejecting the green vegetation I offered her and largely ignoring the available green vegetation during our walks.
3. There must have come a point during Musa's release when the old flora was declining due to absence of the old food and the new flora had not yet become established. At this unsettled moment, an enteropathic bacterium entering her gut would have found an undefended city.

Porcupines normally guard against such intestinal turnovers by a variety of mechanisms. They maintain long fidelity to any chosen food item. Thus, porcupines in both summer and winter will feed on a fixed major food source even though alternate food supplies are available. When porcupines do introduce new food items into their diet, they do so very gradually, sampling small amounts of the new item for several days before committing to it. And once a porcupine commits itself to a major food, it stays with the item for weeks before trying an alternate.

Similar behaviors can be shown for the winter months, when porcupines feed on tree bark. Again, individual animals become feeding specialists, while the species as a whole samples a more diverse diet. Such feeding specializations are best explained in terms of constraints imposed by the gut microflora.

In treating Musa's diarrhea, I had to address not only the issue of fluid and electrolyte loss but equally the issue of reconstituting an appropriate normal flora of the gut. A number of veterinarians and wildlife rehabilitators I contacted suggested a technique called "poop soup." This involves dispersing a dropping from a normal porcupine in water and smearing it over Musa's food. The droppings consist of a mixture of plant fiber and excreted normal-flora bacteria.

When I try the treatment, Musa has reservations. She consumes a small piece of toast smeared with poop soup, then refuses further offerings. I try to fool her by spreading peanut butter over the bacterial smear, but she learns to scrape off the peanut butter layer and leave the rest. Nevertheless, she consumes enough to serve as an inoculum, and begins to produce normal hard droppings again after a 4-week disturbance but only a week after first contact with poop soup.

Signals in the Woods

With Musa fully recovered and beginning to put on weight again, we resume our trips to the Catskills and walks in the fall woods. Her energy, which was vanishingly low during the period of diarrhea, has returned to normal and sometimes higher than normal. At times I find her moving upside down along the ceiling of her wire cage.

On an early-October morning, Musa and I head up Vly Mountain. We will start our walk higher up the mountain than we have in the past, in mature woods frequented by other porcupines. The morning sun ignites reds and yellows all around us, but leaves are falling steadily and a light layer of fallen leaves already paints the ground.

I release Musa at a small opening in the woods, where a marked trail starts uphill toward the ridge. Musa ignores the trail and starts a gentle ascent of her own, moving uphill from the isocline. Fifteen minutes after starting her climb, she carefully investigates the ground under a red maple tree. She sniffs twigs, leaves, and the ground. The maple is about 15 feet downhill from a multitrunk linden, and that is where she moves next.

This is the jackpot! She sniffs endlessly and excitedly all over the area below the tree, focusing my attention on it as well. A resident porcupine has been feeding in the tree; the ground is strewn with niptwigs, the nipped-off terminal branches that porcupines cut from the tree in order to consume its leaves. Some of the niptwigs are fresh, indicating recent feeding.

I also find, not far from the tree base, two piles of fresh porcupine droppings. One pile consists of small-size droppings, the other, in close proximity, of droppings noticeably larger. This is a nursing station, where a mother and her baby have met at night.

The discovery has transformed Musa. After furiously sniffing the ground under the linden, she begins to vocalize: "Mmmmmmmmm . . . mmmmmmmmm." It is a low, extended vocalization that I have never heard Musa give before. It is the voice of a baby porcupine calling its mother! Musa is now possessed. She walks toward me as in a trance, climbs over my boot, and approaches the linden tree. Putting her front paws on the trunk, she prepares to climb. Reluctantly, I stop her—the tree is too large. She backs down but quickly tries again. She makes four attempts to climb, and spends an excited 30 minutes under the tree before moving on.

Musa's behavior has been a revelation. Over years of tracking porcupines, I had come to think of them as solitary animals, sitting alone in trees through long summer days, or hunched in dens, dreaming winter dreams. The only exceptions to this have been mother-baby pairs observed from May to September, and male-female pairs during mating season. Yet Musa's reaction to the presence of other porcupines has been one of highest excitement. This is how a human being marooned alone on an island might react to first contact with another human.

But granted similar emotions in humans and porcupines when making contact, there is an important difference in how humans and porcupines make that contact. Humans respond to one another by sight and sound (language). Porcupines, in more classical mammalian fashion, respond to each other by signals involving the sense of smell; sight and sound are secondary channels.

For olfactory messages, a number of signal channels are open to porcupines. The perineal glands appear to play a major role in mating behavior (Roze et al. in press). The rosette warning odor signals arousal and readiness to fight. But undoubtedly the signal channel most widely used is that of urine. It was the olfactory message left in urine that Musa responded to in our Catskills walk.

If human sight functioned at the same low level as the human sense of smell, all of us would be legally blind. Our osmic impairment makes it hard both to appreciate the importance of this sense in other mammals, and to demonstrate its importance in an unambiguous way. Like the blind man feeling the elephant's trunk in order to decode its shape, we must use other senses to demonstrate the importance of olfaction to porcupines.

One starting point is the observation of porcupines in winter. As the animals travel in the snow between their dens and feeding trees, they leave urine marks that stand out in the snow. Unlike white-tailed deer, which broadcast their urine in snow-melting, irregular showers, porcupines deposit urine in a carefully metered fashion. They deposit it either as small, strung-out droplets, or as a thin stream near an important object. Thus one finds a hatchwork of fine urine streams at the den entrance, and more of the same at the base of the feeding trees. The connecting feeding trails are either unmarked or lightly marked with scattered fine droplets. It is important to realize that porcupines don't simply have poor sphincter control. When an animal goes on a long hike in the snow—for instance when moving to a new den—its snow trail remains pristine, unmarked by a single drop.

What makes these winter observations possible is the visual record left in the snow. No such record is available in snow-free months, but it is not unreasonable to assume porcupines would similarly mark their summer food trees, to which they may return repeatedly. If nothing else, such marking would help them locate the trees in the dark, when the bulk of their feeding takes place.

Another piece of accessory evidence points to the importance of porcupine urine-signaling—their reproductive behavior. During the fall mating season, females sit in trees and advertise receptiveness by means of pheromones released into the urine. The bouquet attracts males, who travel from afar to sit in the tree with the female, with stronger males displacing the weaker ones. All the males can do at first is sit because the female is not yet ready to mate. She will mate only when she ovulates (goes into estrus). The remaining male can speed up estrus by spraying the female with his urine, as described in Chapter 9. The signaling potency of porcupine urine is enhanced by its significant albumin content, which presumably acts as a carrier protein that concentrates and extends signal life.

I get an illustration of this when Musa and I return to the Catskills 1 week later. During the week since her first encounter with her own kind, an inch of rain has fallen. Strong winds have downed branches around our cabin and have uprooted trees higher up the mountain. The fall foliage stands much thinned, with the sugar maple in front of the cabin completely bare. The rainfall, high winds, and freshly fallen leaf litter should give a severe test to any urinary signals left on the ground.

I take Musa for her walk in late afternoon, releasing her slightly higher up the trail than I did a week before. Again, Musa heads north along the isocline. After a while, I notice she has picked up her week-old trail. Along the way, she explores a denlike cavity under the root mass of a freshly windthrown beech tree. Then she continues toward the linden tree where she had her epiphany a week before. She repeats the hard sniffing at the base of the red maple and the ground under its canopy.

She then moves the 15 feet uphill to the linden where a mother and juvenile had fed and nursed a week before. The ground below the canopy remains littered with old niptwigs; no fresh ones are visible. The large linden leaves, still retaining some green, are falling and gradually obscuring the niptwigs. Musa once more is furiously sniffing the ground, but perhaps because the urine signal is now a week old and has been weakened by rain and falling leaves, she does not give her "mmmmmmmm" call. She tries climbing the linden trunk twice, then tries climbing a linden sapling in the multitrunk complex. I frustrate all attempts, and she goes on to try climbing the same nearby red oak she attempted a week before. I stop her again. Eventually, she settles into a crevice under a nearby flat rock and settles down. I igloo her, and we return home.

Release

Next day, a morning drizzle intensifies into a rain, then into the season's first snowfall. Large flakes fall heavily, and for a while the ground turns white. It is now the second half of October, and winter is knocking on the door. I weigh a second release for Musa.

She has made up all of her recent weight loss, and has been adding body mass at a rapid rate. At 4 kg body weight, she now substantially outweighs her wild contemporaries. I think of this as a safety cushion that should tide her over the inevitable weight loss that will follow on release in the winter woods.

Another factor in her favor: if I release her in a porcupine-frequented area, she should be able to take advantage of the collective wisdom of other porcupines. More than that, I now believe that contact with other porcupines should give her an emotional boost over her current position of isolation, as demonstrated by her reactions to urine under the linden tree.

It is this factor that weighs most heavily in my decision. The end result I want to see for Musa is for her to be an independent animal, living among her conspecifics in the wild, not a controlled being accepting the confines of her cage. Although Musa lets me handle her now with little fuss, her first instinct on walks in the woods remains escape.

I am left with one concern—the possibility of renewed diarrhea following release. I prepare for this with the old trick of introducing ground-up droppings of wild porcupines into Musa's diet, to inoculate her with the microorganisms she will need. But Musa has a better idea, as I am to discover shortly.

I release her on a pre-Halloween weekend in the Catskills. Once more, I attach a radiocollar for security, install her in an Igloo, and start up the mountain to her release site. The recent snowfall has melted; all the leaves are now on the ground, and the forest is rich in leaf smells as we slowly make our way uphill. Periodically, I put the Igloo down to have a rest, and to give Musa a chance to orient herself. To help her in this task, I have partially propped the Igloo open with a large woodworker's clip.

After a little pause, I see vibrissae protruding from the Igloo opening, then a little front paw poking out, and then both paws and the tip of the nose, sniffing with excitement. The black body parts stand out sharply against the white lip of the Igloo, and they look delicate and beautifully formed.

We reach our goal, Split Rock den, shortly after noon. Split Rock is a deep, comfortable den that porcupines over the years have prized. It has seldom lain empty, but it appears unoccupied today. There are no fresh droppings near the entrance, and the passageway holds drifted leaves.

Yet Musa informs me that a porcupine has been here previously. She sniffs the ground intently, making occasional "mmmmmmmm" sounds. She finally disappears inside the den, and I hear the rustling of leaves and Musa's soft "mmmm." There must be porcupine smells inside but not an actual porcupine because I do not hear the squawking sounds of a disturbed resident. After 10 minutes, she emerges and gives a thorough sniffing to the ground and to a small rock ledge holding old porcupine droppings.

Then Musa does something truly amazing. She picks up an old dropping, bites off a small piece, and swallows it. With this act, she has prepared herself for feeding on the local vegetation that the previous resident was feeding on. She has inoculated her gut with the microorganisms she will need in this specific location. I now surmise that it was the unavailability of such an inoculum at her first release site (because of the absence of other porcupines) that led to her debilitating brush with diarrhea.

Coprophagy as a mechanism for acquiring essential gut microflora is known in a number of other herbivores, for example, young elephants and horses. This, however, is the first observation of self-inoculation with relevant gut microflora in a porcupine. Again, previous ignorance of this behavior follows from the fact that biologists seldom study porcupines by walking with them in the wild.

Musa continues her careful exploration of the area outside the den entrance. She even comes up to the tip of my shoe and sniffs it without showing any quill reaction. I remain immobile. She also sniffs the small offering of acorns, apples, and rodent chow I have left outside, but touches nothing. Three minutes later, she retreats back into the den. I depart.

Slowly, I make my way down the mountain, an empty Igloo in my hand. There will equally be an empty spot in my heart—the walks in Alley Park, the trips back and forth to the Catskills, the periodic negotiation of trust when I pick her up, and also the periodic disagreements that are soon patched up—all will now become a history. But I also know that she will experience a larger life on her own, and with her own kind, than she would have had in my protective, and constricting, shadow.

That evening, a heavy rain starts to fall, and continues all night. By the following morning, over 2 inches show in the rain gauge and the rain is continuing. I load rain gear and radiotracking equipment, plus some apples and acorns for Musa, and start up the mountain to look for her.

Belatedly, I remember an outstanding defect of Split Rock den—it is not waterproof. The den is deep, dry, and snug in the winter, when an insulating layer of snow seals all crevices. But every spring the snow layer melts and meltwater percolates through a crevice in the roof and sends the resident porcupine searching for a new den. A substitute, which I have named Dry den, is available less than 50 feet away. It never floods, and porcupines over the years have left a thick layer of droppings as evidence of their occupancy. But none of the residents stay long, because Dry den is shallow and offers poor protection against the cold. I wonder if this is where I will find Musa.

As I make my way slowly up White trail, I find the trail blocked by two freshly downed beech trees—evidence of the storm's power. On reaching Split Rock den, I find the acorns and one of the two apples gone, but there is no radio signal from Musa. In the earphones, I hear only static. This means Musa is neither in Split Rock den nor Dry den, and probably in no other den on this side of the mountain. The complete loss of signal suggests she has crossed over the ridge and is now somewhere on the other side. I climb to the ridge top and listen again. At once, the signal comes in strongly, mostly in the inactive mode but with short bursts of activity, as when a sleeping animal stirs momentarily.

The downhill passage from this spot is a difficult one. Striped maple grows so thickly I must move sideways at times. I spend an hour in the steep and slippery terrain before I can reach Musa. Her surroundings are forbidding. A short distance downhill, the trees have been thinned by owners of a hunting club. A central deer blind commands long lines of fire, and I am thankful that she would not be moving during daylight.

Musa is resting in the porcupine equivalent of a lean-to. Instead of the comfortable Split Rock den or Dry den, both used by porcupines for years, she is wedged under a slanting rock outcrop that does not qualify as a den. The crevice she is sheltering in is so shallow that her back is in full view, and her tail must be getting wet. When I arrive, her radio clicks into the active mode and her quills rise erect. In spite of her uncomfortable quarters, she has no intention of coming back with me. I take some photos, leave her a handful of acorns, and wish her well. I head home through bare woods, the last of the fall leaves stripped from the trees and restitched together on the ground in a continuous, warm-colored quilt.

I return weekly or more frequently to track Musa. I find her settling into permanent winter quarters near the top of the ridge, but on the north slope, opposite that of Split Rock den. On November 10, 2 weeks following release, I manage a capture.

Musa is in a den under a great glacial erratic. She is in porcupine country; a layer of old droppings covers the den recess, and witched trees on the slope below testify to years of porcupine winter feeding. I find Musa's own feeding sign—six young striped maple seedlings, with bark stripped from the lower trunks. The feeding sign has "juvenile" written all over it. Much as a human baby might leave a messy table and plate, Musa's bark-feeding is undisciplined. More feeding attempts have been initiated than completed, and the one tree with extensive bark removal shows a feeding scar with messy, irregular edges and a chaotic pattern of bark-scraping. Another juvenile sign is the tree selection. An adult porcupine would have chosen a larger tree, and fed in the crown instead of on the trunk at ground level. But Musa's feeding habits are not unusual. Other porcupines her age would show a similar pattern. The real test of feeding skill is physical condition. How is she doing on her own?

The capture proves easy. Musa's den is shallow, and she keeps her quills only partly erected. Her weight is 3.4 kg, a loss of less than one-sixth of her prerelease body weight. Winter weight loss is something that all porcupines experience, and Musa's weight remains above average for her age group. The perineal area shows no signs of diarrhea. I smile—the little animal is fighting a winning battle. She has teachers and allies—the subtle messages left by other porcupines all around her.

I meet her only one more time (Figure 13.7). It is the weekend before Thanksgiving, and the day is sunny and exceptionally mild. Musa is where she was before, just below the mountain ridge on the north face. The warm day has called her out of her den—I find her up a small beech tree, whose upper branches display several fresh feeding scars. She sits comfortably in the twisting branches of the beech tree, and all around her are signs of old porcupine occupation: tree crowns witched by repeated feedings and rock outcrops offering deep den sites. Far below us, Little Westkill Valley and the distant Schoharie glow through leafless branches. I reflect how often good porcupine habitat includes such lovely vistas. Musa is finally home.

Figure 13.7. Final view of Musa, below a ridge of Vly Mountain. (Photo by UR)

She is aware of my presence; she has stopped feeding and watches me silently from her perch. Her quills remain down. I violate this trust by reaching up to her low branch and capturing her. Capture, weighing, and release take less than a minute. Her weight today is 3.05 kg, down from her previous weight but still squarely in the middle of the range of wild porcupines in her age class. She retains considerable reserves for overwinter survival.

This meeting on a sunlit afternoon turns out to be our last. Thanksgiving and other matters keep me in the city for the next 3 weeks. When I return, fresh snow covers the ground, and the deer-hunting season is on. I put on my bright orange cap and go radiotracking. Along the woodland trails I find tracks of coyote, fox, and grouse, but surprisingly no deer. Then I find the fresh tire marks of hunters' ATV's and understand why deer are traveling elsewhere. Musa's signal pulls me upward to the same high north face as before, but I notice one difference—the signal never cycles into the active mode. Even a sleeping animal should occasionally stir or scratch or change position. I wind up at a rock outcrop only 20 feet below the ridge, and the signal is so strong it's booming. I see an opening that might be a porcupine den, but no porcupine tracks emerge into the snow outside. And ominously, the signal remains in the inactive mode.

I crawl inside and see scattered porcupine droppings, but no animal and no collar. Finally I retreat, take off my receiver and antenna bag, and plunge in again to get as deep as possible. My flashlight reveals a more substantial accumulation of droppings, obscured in part by windblown leaves. And finally I spot the radiocollar, wedged in among the leaves where Musa has shed it as her neck diameter has shrunk. She is now fully on her own—my gift of apples and acorns cannot be delivered.

Over the years, many of my porcupines have dropped radiocollars. We are often reunited when they descend the mountain in the spring to seek a salt supply, which I control for upper Roarback Valley by virtue of my salthouse. I therefore don't abandon hope of meeting Musa again.

The winter passes, and I get a string of salt-hungry porcupines visiting in the spring and summer. Yet none of them is a young animal with an asymmetrically healed lip. I think of the possibilities. Porcupines do not generally cross a mountain ridge to explore the opposite valley. Of course, there are darker scenarios, but I want to imagine Musa making her way down the north face, exploring the Little Westkill valley that looked so sparkling through November trees high above, and returning home with a bellyful of salt.

If that is where Musa is now, she will have spent, for a porcupine, a cosmopolitan childhood: the dangerous traffic of the Notch, a Queens backyard, a spacious city park, and periodic visits to a Catskill mountainside. Perhaps each place taught her something new. But equally true, in each place she told me something about herself: the difficult feat of learning to climb a tree, the subtleties of navigating in the woods, the uses of geophagy in taming a wild-plant diet, the urine signals of her conspecifics, and the coprophagic route to an appropriate intestinal microflora.

The result has been a process similar to that of friendship: the slow opening of inner secrets that add dimension and empathy to a life not one's own.

14 The Oldest Porcupine

In 1969, my wife and I bought 67.5 acres of wooded land on Vly Mountain, in the northern Catskills. We signed a mortgage and recorded a deed affirming ownership of the roadless mountainside, part of a farm established by Daniel Angle in 1807 and abandoned in the 1950s. Unknown to us, another owner took possession of our acreage in 1980. She made no mortgage payments, filed no title with the county clerk, built no cabin. Yet she remained in residence for 21 years, dying in extreme old age. She was a porcupine whom I first captured in winter 1981–1982 outside a massive rock den high on the mountain. I named her Squirrel in recognition of her alert, optimistic features and bright gaze (Figure 14.1).

Since our first encounter outside her winter den, I have spotted her in her tree, den, or salt source an estimated 450 times, and have physically captured and examined her more than 60 times. At our final meeting in March 2001, deep snow covered the ground as it had at our first meeting 20 years previously. Squirrel sat in a far more flimsy den inside a hollow beech tree on the eastern edge of a hemlock grove. A large opening in the side provided a full view of her head and upper body, and of the radio transmitter below her chin. Holding my camera in front of me, I slowly moved closer, capturing photos until I was only 2 feet away. Squirrel continued facing me, her eyes unblinking, in a behavior very unusual for a wild porcupine. Had she finally accepted me, her long-term co-resident, or was her gaze already focused on a more distant prospect? I left two Red Delicious apples (her favorites) at the den entrance and departed.

When I returned 2 weeks later, additional snow had fallen and my snowshoes carried me high off the ground. At the cabin, Squirrel's radio signal was coming from the previous direction, but ominously, its beat stayed invariant and in the inactive mode. I set out at once and reached the hollow-beech den. There were no signs of activity—no trail in the snow from feeding tree to den,

Figure 14.1. Squirrel at age 10, wearing a radiocollar. (Photo by UR)

no hemlock boughs dropped from the canopy. The den was empty, yet when I stepped back and made a sweep with the antenna, the signal seemed to issue directly from the hollow beech. I unbuckled a snowshoe and, using it as a shovel, began to excavate the den entrance. White, loosely packed snow soon gave way to a crusted layer, with scattered porcupine droppings and orange urine stains, but no body.

I stepped back and searched around the beech in a wider circle, finding carnivore tracks in the clean upper snow. The footprints showed four toes, not five—therefore a canid, not a mustelid. The prints were small, therefore a fox, not a coyote. But the tracks, having approached the tree, departed again without sign of further exploration. A porcupine is not food for a fox even after it can no longer defend itself.

Now I turned my attention to the back of the den tree. The snowshoe almost immediately excavated a dark mass—a bristling of tail quills. I dug further and found Squirrel, lying on her side just outside the den, with left foot and arm tucked under her body. There were no signs of injury. The body gave an impression of health, with fur dark, clean, and puffed up, the guard hairs tipped in white. I carried her back to the cabin like a child, slung over my back, and felt the heaviness of her body. The balance confirmed it—she weighed 7.2 kg, a high winter weight for any porcupine and very high for her age.

Longevity in Porcupines

Squirrel is not the oldest porcupine I have met—that distinction belongs to a female I encountered in April 1988. I had climbed the north face of Vly Mountain and descended far on the other side, where a little stream ran in a spring-greening valley. I had been led there by the radio signal of a dispersing female, Chili. Suddenly a small porcupine materialized from brushy ground and hurried across my path, heading for the stream. Its muzzle was eerily white. Because I had been tracking Chili, I carried my capture gear of that time: a wire cone, heavy rubberized gloves, and anesthetic. As I ran to head off the porcupine, it charged blindly, hitting my boots. I scooped it up in my wire cone, anesthetized it, and examined it close up. The animal was medium-size, weighing 5.25 kg, sex female. The white of her muzzle extended around her eyes. The grinding surfaces of her cheek teeth had been worn smooth.

In grazers such as sheep, the loss of cheek-teeth grinding surfaces leads to starvation and death. Porcupine tooth grinding surfaces last longer because porcupine diets do not include the abrasive, silica-rich grasses of grazing animals. Yet efficient digestion of a porcupine's winter diet—the inner bark of trees—requires considerable grinding and milling, and I wondered how this old animal managed in winter months. I have never seen her again, and have no independent measure of her age. Based on comparisons with the 21-year--old Squirrel, whose molars remained serviceable though clearly worn, and whose muzzle retained its dark hue, I estimate the old female's age at capture at 25–30 years.

How long can a porcupine live? The most careful published study, by R. D. Earle and K. R. Kramm in 1980, uses the technique of tooth-sectioning in dead animals. Moliform teeth, following decalcification and staining, show growth rings corresponding to years since tooth eruption. The authors collected skulls from 247 culled animals, and displayed the results of a subset of 74. The median age of the population was 5 years; the oldest of the 74 had been killed at age 18.

The data of Earle and Kramm show a drop-off in animal weights with advanced age; body weights begin to drop after age 12, from highs of about 7.5 kg at ages 4–12 to approximately 5.5 kg at age 18. For comparison, Squirrel's final weight of 7.2 kg was near her lifetime high. As I note later, at least part of her high weight may reflect the absence of the twin drains of pregnancy and nursing. Of the 198 individual porcupines encountered in my long-term study, Squirrel remains the oldest with a life history.

Longevity Factors: Extrinsic Mortality

In the animal world, several factors promote longevity. One is body size; large mammals live, on average, longer than small ones. Thus, one bowhead whale killed by Inupiat Eskimos in northern Alaska was determined to be 211 years old (Rogers 2000). A bottlenose dolphin, a smaller animal, may live only to 48 years, averaging about 20 (Duffield and Wells 1990). There are no mouse-size whales, but if they did exist, they would be expected to have mouse-size life spans of 1–2 years.

Among rodents, porcupines approach the upper limit of body size—the African hystricids are, in fact, the largest rodents on that continent. It is therefore not surprising that the family Hystricidae (Old World porcupines) have the longest life expectancies among rodents. The Erethizontidae (New World porcupines) follow closely behind (Dammann 2006).

But body size is linked to a more fundamental biological parameter—extrinsic mortality. Large animals are generally less vulnerable to predators and less subject to fluctuations in weather and food supply than small ones. (However, a number of small animals with protected lifestyles can also achieve great longevity. Examples among mammals include the naked mole rats; among insects, the queens of several ant species.) The link between extrinsic mortality and longevity is rooted in evolutionary theory. If extrinsic mortality is high for a given species, hardly any individuals survive to old age. This means evolutionary selection for longevity is too weak to oppose the accumulation of germ-line mutations with late-acting deleterious effects (Kirkwood and Austad 2000). This is the mutation-accumulation theory of aging, first proposed by Medawar in 1952.

A second, related theory is that of antagonistic pleiotropy (genes with multiple effects are called pleiotropic) (Williams 1957). According to this theory, genes with good early effects should be favored under conditions of high extrinsic mortality even if the same genes have negative effects at later ages. This will be true even if the early, beneficial effect is small and the late deleterious effect a large one.

A third theory of aging, which likewise recognizes the link to extrinsic mortality, is called the disposable soma theory (Kirkwood 1977). This theory focuses on the trade-off between investment in reproduction and investment in a well-maintained, long-lived body. If the odds of surviving to old age are low, a greater evolutionary payoff will be realized by early and abundant reproduction. Some organisms (semelparous organisms) carry this strategy to its extreme—they reproduce and die immediately afterward. Animal examples include mayflies, many moths, lampreys and eels, and Pacific salmon.

A porcupine follows the opposite strategy. A female does not begin to mate until her second year of life and gives birth at the beginning of her third year. This was true of Squirrel in summer 1982, when my relentless searching revealed no baby. (Porcupine mothers hide their babies and meet them once a night for nursing; see Chapter 9.) To make absolutely sure, two students and I organized 36 hours of round-the-clock observation. During the interval, Squirrel moved from resting tree to feeding tree and back, and gnawed a fresh deer bone for her sodium supply, but never met with a baby. It was not until her third year, in 1983, that she became a mother.

All three evolutionary theories of aging (the mutation accumulation theory, the pleiotropic theory, and the disposable soma theory) focus on the same central factor—the degree of extrinsic mortality. Porcupines shine as defensive specialists. Although their antipredator capabilities may break down with specialized predators like fishers and mountain lions, they can more than hold their own against such powerful carnivores as coyotes, which at the time of this writing have driven the local deer population out of sight.

Porcupines are likewise well-defended against most parasites and disease organisms. The only parasite with serious impact is a recent addition—the scabies mite. And even there, the porcupine is relatively protected by virtue of its solitary lifestyle; an infected animal may die without infecting another porcupine.

Porcupines are well-defended against winter cold. Their fur and quills offer excellent insulation, and cold protection is enhanced by their use of a winter den, which may be 10–20 °F warmer than the outside air. Their food, in the form of inner bark of trees, is never far away, necessitating only short feeding trips in the snow.

Because of such interlocking and mutually reinforcing body traits and behaviors, the porcupine is well suited for the evolution of longevity. What these Methuselah factors may be at the molecular level—DNA proofreading and repair mechanisms, free radical and reactive oxygen defenses, telomeric enhancements, tumor suppressor genes, or still others—remains to be discovered. But like human babies, most porcupine newborns can reasonably expect a long and fruitful life stretching ahead at birth.

Testosterone

There is one important longevity factor that not all porcupine newborns can share—being female. As with their human counterparts, female porcupines live longer than males. In 2004, the life expectancy at birth for U.S. human females was 80.4 years; for males, the figure was 75.2 years (Miniano et al. 2007). No one has collected comparable data for porcupines, but available data all point to a female advantage. For example, Hale and Fuller (1996) calculated 1-year survival probabilities for adults in their study population in Massachusetts. For males, the figure was 0.56, for females 0.74. In my own 26-year telemetry study in the Catskills, adult females remained under observation an average of 4.14 years ($n=42$) and adult males 3.17 years ($n=28$). These are not true longevity measures; adults typically entered the study at ages older than 1 and could drop out for reasons other than death. Nevertheless, they do suggest longer porcupine life spans for females than for males.

Why do females outlive males? An important male–female difference is body testosterone level. While this hormone is present in both sexes and in humans is responsible for such important female behaviors as sex drive, its level is much higher in males. In males, it is responsible both for sex drive and for the development of male secondary sexual characteristics. In male porcupines, these include a conspicuous tuft of white quills on top of the head and the descent of testes into reproductive position.

Testosterone shortens life span in human males, as shown by the fact that castrated males outlive intact ones by up to 15 years (Hamilton and Mestler 1969). The life-prolonging effect is inversely proportional to the age at which the operation was performed.

Part of the testosterone effect follows from male aggression. Human males are more likely than females to die in homicides and accidents. But in developed countries, males are also twice as likely as females to die of infections and parasites (Owens 2002). In Kazakhstan, with a less well-developed health-care system, the difference is fourfold. What would it be for porcupines, who receive no health care whatsoever?

Testosterone is an immune system suppressant (Zuk and McKean 1996; Duffy et al. 2000). Tests in starlings show that physiological levels of testosterone suppress both antibody production (humoral immunity) and cellular immunity. The mechanism of such immune suppression is not known. One possible mechanism may be that testosterone induces a stress response. Physiological levels of testosterone raise cortisone levels almost 100% in males and to a lower degree in females (Duffy et al. 2000).

The male–female differences in stress response are very much on display in porcupines during the capture process. My current capture technique involves running up to a porcupine, clapping an Igloo cooler over the top of the animal, and closing the top. With few exceptions, I can quickly tell whether my captive is a male or female. Males throw themselves about, try to chew their way out through the plastic walls, and give every indication of experiencing high stress. By contrast, females sit quietly, their heads pushed down in protective position, and wait for me to open the top before erecting their quills and defending themselves.

This brings us back to the discussion of evolutionary trade-offs. The energy that males invest in a stress response and in the high-energy tasks of mating and intermale competition must come from somewhere. One place to borrow from is the energetically demanding task of maintaining an immune system—a system that works to promote health and long life.

Motherhood

Squirrel, happily free of the testosterone tax, began her life in spring 1981. At the end of her first summer of life, she presumably followed the female imperative of her species: she left her natal territory forever. When I first encountered her in winter 1981, she was a new arrival in northwestern Roarback Valley. She would spend the rest of her long life on her chosen hillside, learning its food trees and rest trees, salt sources and den sites, forest trails and openings. And here she would conceive and bring into the world her lifetime 13 offspring.

Squirrel's first baby, born in spring 1983, was the female Bus seen in Color Plate 8, whose short life was ended by a hunter's bullet in winter 1983 (Chapter 2). Over the years, Squirrel produced 5 males, 5 females, and 3 babies with sex undetermined because they could not be captured for examination (Table 14.1).

Squirrel's maternal skills improved with maturity. Although her first daughter was killed by a hunter, it is unclear that she would have survived her first winter on her own. A juvenile's first-winter survival depends crucially on its fat stores, as reflected in body weight. In the fall and winter of 1983, Bus never reached more than 1.9 kg in body weight. By contrast, the offspring of 1987–1992 had already passed this milestone by July of their birth year. Even Crew, her last offspring, had reached 1.8 kg body weight by July of his first year (Table 14.2).

Table 14.1. Squirrel's lifetime reproductive output

Female offspring	Male offspring	Offspring, unknown sex	No offspring	No data
1983	1985	1984	1982	1995
1986	1988	1990	1996	
1987	1989	1994	1998	
1992	1991		1999	
1993	1997		2000	
			2001	

I next recount the story of Crew, Squirrel's final offspring, born in the same woods as his oldest sister 14 years previously and whose ending in its own ways was as tragic as hers.

I begin the year 1997 without a radio signal from Squirrel. As a concession to her advancing age, I had removed her radiocollar in July 1993 and substituted a plain black nylon band to help me recognize her at a distance. Once a porcupine is radio-silent, the chances of encountering it drop drastically. But I knew enough of her favorite trees (Figure 14.2), dens, and salt sources to capture and examine her 13 times in the intervening period and

Table 14.2. First-year weight gain in Squirrel's offspring

			Weight (kg)					
Year	Sex	ID	July	August	September	October	November	December
1983	F	Bus	1.1	1.5	1.7	1.6	1.7	1.9
1987	F	Eft	2.2	2.6				3.3
1988	M	Quill	2.0		3.3		3.1	
1989	M	775		2.2	3.9			
1991	M	779	2.2	2.6				
1992	F	S92		2.5				
1997	M	Crew	1.8					

Notes: ID, identification; F, female; M, male

Figure 14.2. Squirrel feeding in the spring. (Photo by UR)

keep track of her annual lactation status, weight, and general condition. Squirrel had produced a daughter in 1993 and another baby (sex undetermined) in 1994. There were no data for 1995, but clear evidence against motherhood in 1996 when two midsummer capture exams showed no lactation.

Most puzzling to me is the intrusion of other female porcupines into the core area of Squirrel's old territory. In 1996, both Palin and Heart had spent extended periods of time in some of Squirrel's favorite linden trees, and both were mothers that year. In past years, no foreign female had made more than a brief foray into Squirrel's territory. Is Squirrel getting too old to evict intruders? Is her body no longer sending out the appropriate chemical signals? I decide that, given the opportunity, I will re-fit Squirrel with a radiocollar to learn how she deals with intruders and to see whether she still inhabits the same core area as in past years.

The opportunity comes on June 14, 1997. After a day of radiotracking and strenuous outdoor activity, I feel tired as I head for the salthouse at midnight. I hope to capture porcupines during their spring and early summer salt binge, when animals come from distant spots to sample the salt supply I have provided. I carry two Igloo plastic ice chests with closable tops—I am prepared for more than one animal.

But when I reach the salthouse, I find it abandoned. I now switch to plan B; I set up a folding cot inside the salthouse, shut off my headlamp, and lie down to wait for porcupines. I fall asleep immediately.

A deep sound awakens me at 1 a.m. I feel disoriented at first, inhabiting a dream. Am I by a lake? What is the strange sound? Then reality informs me—I am inside the salthouse, there is a porcupine outside. Silently I slip on my boots, pick up an Igloo, adjust the headlamp, and creep to the door. The deep rumblings continue, with interruptions as the porcupine moves to attack a new saltstick. I open the door, turn on my headlamp, and confront a large porcupine with a black nylon collar—Squirrel! I igloo her with difficulty as she clings to her saltstick.

She stops fighting once inside the Igloo. I carry her to the cabin for a physical exam. A black "S" tattooed in the groin confirms the identification. Squirrel's weight is 7.2 kg, higher than any of my other porcupines of the year. I notice how large her footpads are—the size expected in large males. She is now showing signs of age—her chin is gray, and gray circles have formed around her eyes. But her premolars and molars remain fully ridged, and she is lactating—she has a baby in the woods. I attach a radiocollar to follow this old mother, and to see how she deals with the younger females who have invaded her territory.

Squirrel leads me to her baby the very next week. I track her to the top of a leaning linden tree, overarching the Roarback at Porcupine Crossing. This is a tree much favored by Rebecca in years past. It lies on the border between Squirrel's and Rebecca's old territories.

A linden sprouting from the same root mass has a hollow base, reminiscent of linden tree 766, where Squirrel once hid Bus. I look inside the opening and see nothing, yet the site remains suspicious. Squirrel has not chosen her tree to feed in—though there are niptwigs on the ground, all are several weeks old. And the floor of the hollow linden trunk shows signs of recent traffic—scrape marks and fresh quills.

I move to the side, where there is a small windowlike hole in the trunk. Pressed against the opening is a thatch of black and white quills—Squirrel's baby, sitting on a platform inside the trunk! My day feels complete. I leave the baby in its hiding place, confident we will meet again soon.

I have to track one of my males, who has moved so far north and east I must search for him by car, and thus I find myself on County Road 23C, high above the Roarback and Schoharie valleys. I leave the car by a green field and look back at Roarback valley and Squirrel's distant resting tree. The Lexington panorama opens before me: the Notch between Vly range and the Westkill range, Saddleback in the foreground. The fields are crowded with summer's high grasses and field flowers. The mountains stand deep green, with occasional tracery of roads and habitation. The sky is busy with clouds, and the sun pours one more charge of photons to speed the growth of this rich land.

I look at the scene with feeling. The unfolding of tree leaves is now at its maximum, and plant and animal life are racing before approaching autumn. I suspect that Squirrel and I are sensing the same turning of the earth, she from inside her green canopy, and I from the high ground above her. We are both part of a landscape that will survive without us, but whose beauty gives our lives their definition and warmth.

Over the course of the next 2 weeks, Squirrel reestablishes herself in her core territory, while Heart and Loretta, two females who had intruded on her space earlier this summer, are displaced. Heart now occupies adjacent space to the north, Loretta adjacent space to the south, and Palin is found to the south of Loretta. All females are nursing their young of the year. I find no evidence that Squirrel's reoccupation of her territory is accomplished by direct contest. None of the trees used by the displaced females shows an unusual quill count on the ground below to indicate fighting.

By contrast, a confrontation of years past between Squirrel and an intruder had been settled differently. When an outside female intruded on Squirrel's territory, I found evidence of battle—70 quills on the ground below Squirrel's rest tree. The foreign female disappeared, and no further quill clusters could be found under Squirrel's trees.

In 1997, my expectation of capturing the baby is finally realized on July 19. I go to the salthouse at midnight and find it abandoned. On a hunch, I check the sweet apple tree in the orchard and find Squirrel feeding in a topmost branch, out of reach. I leave her alone and retire to the salthouse. I am awakened at 3 a.m. by the night chill. Silently I put on my boots and prepare to leave when I hear arriving porcupines. There is some grunting and then the deep music of porcupines rasping wood. The sounds come from two directions as two porcupines work outside. I emerge silently, switch on my headlamp, and see the two animals. Squirrel with her radiocollar is on my left, her baby on my right, continuing to chew a saltstick even as I watch. I have only a single Igloo on hand and use it to capture Squirrel. I then use my rubber gloves to capture the baby, who has not moved. I carry both back to the cabin, where I install them in adjacent cages, with apples as peace offerings.

From past experience, tonight's double capture of mother and baby at the salthouse is an unusual event. Mothers seldom bring their young of the year to the salthouse, presumably because babies obtain their necessary sodium from the mother's milk.

In the morning, both have consumed their food, with the baby leaving messy scraps on the floor. During my physical exam, I find the baby is a male, weighing a hearty 1.76 kg. At the same time in 1983, Squirrel's first baby had weighed only 1.06 kg and did not reach 1.75 kg until November 11. Squirrel herself has more body mass today (7.23 kg) than she had in midsummer 1983 (5.6 kg). Body reserves are important in generating an adequate milk supply.

As I study the baby, I note a flat top to the quills and guard hairs above his head, suggesting a crew cut; I therefore name him Crew. I attach a small radiocollar, and after a 1-hour recovery period release him together with his mother. The two head off in opposite directions, but at night I find them together again in the sweet apple tree in the orchard, Squirrel's lower-frequency radio signals intermixing with Crew's high cheeps.

Mother and son are together again the following afternoon, resting in a large linden tree. Squirrel lies high in the canopy, her son on a stout branch below hers. Perhaps my presence discomforts the little animal, who gets up, walks to the trunk with back held high, and descends out of sight.

A week later, I encounter the pair by accident. I am passing through the orchard at 1 a.m. after leaving the large male Evander. A commotion in the grass catches my attention. As I approach, I realize Squirrel has been nursing Crew on the ground near the sweet apple tree. I also note that Squirrel is wearing her radiocollar, while Crew is bare-necked. I had attached his collar loosely to give him room for growth. I capture the baby easily and keep it overnight for recollaring.

In the morning, I quickly recover the collar in open woods near the orchard. Luckily, it is lying on the ground, unlike other collars lost in treetops or in the depths of rock dens.

I reattach the collar and examine Crew. He now weighs 1.83 kg, a healthy 1-week gain. He is free of ticks, which plague all porcupines spending time on the ground. With his bright eyes and spiky hairdo, he seems the picture of life and shining promise. But the trajectory of his life is now near its high point. He will fall, dying little more than a month later.

In subsequent weeks, Squirrel and Crew travel everywhere together. I find them feeding at night in the small wild apples of Roarback field, Squirrel on a low branch at eye level and Crew higher up, immobilized in my light. The pair is easily capturable, but I leave them in peace.

A week later, while Squirrel feeds in a linden tree off the lumber road, Crew takes refuge in a small adjacent sugar maple. The ground shows evidence of time spent together, a scatter of adult-size and baby-size porcupine droppings.

My final sighting of Squirrel and Crew together comes on the last day of August. We have visitors for the weekend, and we move with the slowness of a group to visit all the radiocollared porcupines on the mountain. Squirrel and Crew are closest to the cabin—we find them up a large bigtooth aspen a short distance from their previous week's position. Bigtooth aspen leaves, like linden leaves, are an important part of the porcupine summer diet, and today's tree bears a label identifying it as a tree used by Squirrel 10 years previously. The tree trunk is tall and massive, showing little taper, and hints at the high cost of lifting the canopy branches into the sun. For the aspen, this is not an optional expense. Its leaves die in the shade, unlike the leaves of beech or sugar maple.

Squirrel and Crew are sleeping on separate branches and have not yet begun to feed, as shown by absence of niptwigs on the ground. I pass around a pair of binoculars and everyone admires the small tuft of quills and guard hairs high up in the tree, his mother just above him. Then we pass on.

For the next several weeks, I follow Squirrel as she moves very gradually uphill toward the western boundary of her territory. I do not find Crew because of a radiocollar malfunction; a power-saving circuit has silenced the transmitter during the morning and noon hours when I have been visiting Squirrel.

I become disquieted when, at the end of September, I surprise Squirrel on the ground, feeding on raspberry canes. I capture and examine her, carrying her back to the cabin for the exam. The most striking finding is that she has stopped lactating. This might be accounted for by the beginning of the mating season. But her vaginal closure membrane shows no change from the summer condition. And she seems to have stopped lactating some time ago, as shown by the absence of milk-stained fur behind the nipples.

The following week, I wait till Crew's signal comes on in the afternoon and then set out to locate him. A bright autumn light ignites the reds and oranges of maples, and plays them against the remaining greens of oaks, aspens, and lindens. Despite the color change, these woods are familiar. The signal pulls me back to the same bigtooth aspen where I last saw Crew and his mother, but this time the signal is coming from the ground. I stare at the spot, and see a matted mass of quills—the dead body of Crew.

In the stunning realization of death, at first nothing makes sense. How could this young animal, so full of life and health at our last meeting, have suddenly ceased to exist? Finding Crew on the ground is very different from finding a roadkill that has no previous history. Death is not something concrete—it is the absence of life. Death can be experienced only by the living and then emotionally only when the living have shared a history with the dead.

The human mind seeks explanations: What led to the disaster? Was Crew attacked by a predator? His small frame would have been no match for a prowling fisher. Did he fall out of a tree? Did he die of disease? I return to the site three times, gathering a widening net of evidence, until all the pieces fit and a clear story emerges.

Crew was already dead when I captured Squirrel last week. I had passed within a short distance of the spot, and the receiver indicated a strong signal, but I was carrying Squirrel in my arms and the light was beginning to fade.

I eliminate the predator hypothesis when I study the body. The quills are all in one place, not scattered as in a predator attack. There are no missing body parts. The only sign of feeding is on the tips of the ribs—the sign of mice.

I also eliminate the disease hypothesis, which would require the body to lie in a more or less natural position, perhaps in a sheltered spot, and require an undamaged skeleton.

In fact, the body lies contorted, with the neck bent sharply forward. There is damage to the skull—the right zygomatic arch (cheekbone) is separated from the skull. I recover the missing piece—it has been broken off, not chewed off. The right angular process of the mandible (lower jaw) has been cracked and pushed in. The right temporal bone (portion of skull that anchors the cheekbone) is missing a chip. The chipped edge again is consistent with trauma, not chewing by an animal.

All this suggests a fall from a tree. Crew fell from a great height and must have died instantly, either from a broken neck or from the skull damage. The tree he fell from is the very one where I saw him last—the bigtooth aspen carrying Squirrel's 10-year-old label. There is evidence that mother and son fed extensively in the tree; on the ground below lie 28 niptwigs. A porcupine feeds by this technique to avoid putting its full body weight on the small outer branches of a tree. But the branches of bigtooth aspen are notoriously brittle. Crew had ventured out too far on a terminal branch. I see the main branch directly overhead.

It is my daughter Rachel who uncovers the final segment of the tragedy. She points out a flat rock 2 feet uphill from the body, measuring about 18 inches square. On the rock is a pile of 20 porcupine droppings, greenish (signifying leaf-feeding), adult-size, and now in the process of disintegration. They are Squirrel's droppings. After her son's fall, she descended to the ground and waited next to his body for hours. Her mother's devotion had been expressed in this final vigil, and no one can know what went through her mind during those dark hours.

As in a Breughel painting, Crew's death registers in a small corner of the scene. Surviving leaves of bigtooth aspen glow green against the scarlet and orange of a neighboring maple. High overhead, the first undulating waves of Canada geese are flying south. The autumn sunlight warms the air and picks out the yellowing white of a hayscented fern. Only a sharp eye picks out the darker mass of quills nearby.

Menopause

Crew was the product of Squirrel's old age, the last offspring she produced. During the bright fall of 1997, I find her sitting in an ash tree and later, repeatedly, a red oak, unvisited by a male. The reason suggests itself the following winter, when I capture her at Split Rock den in January 1998. She weighs 6.9 kg, a surprisingly high weight for so late in the season, reflecting her abbreviated period of nursing the preceding year. The most surprising finding is an open vaginal closure membrane. This means she has either mated recently or is getting ready to mate; either possibility raises problems.

If Squirrel has in fact just mated, her baby would be born in 7 months, in mid-July. This would be too late for it to put on the body weight needed to survive the following winter. If, on the other hand, she is just entering estrus and becoming receptive to males, she will find no males to respond to her signal. Both Squirrel and potential males are denned up, out of pheromonal signaling range.

I conclude that Squirrel's reproductive system has lost the synchronization needed to integrate into the common mating system. In the Catskills, the porcupine mating season peaks around October 1. In fact, Squirrel produced no offspring for the last 3.5 years of her life. Squirrel had entered menopause, and she had done this under natural conditions, without medical care or predator protection such as might be available under zoo conditions.

In natural populations, menopause has been described in short-finned pilot whales as well as other whale species, ringed seals, African elephants, chimpanzees, the Virginia opossum (vom Saal et al. 1994), and African lions and olive baboons (Packer et al. 1998). This is its first description in the North American porcupine. With respect to human menopause, up until recently the biological rationalization has focused on the "grandmother effect"—females past a certain age can maximize fitness not by giving birth to additional children but by investing in the upbringing of their grandchildren (Shanley and Kirkwood 2001; Hawkes 2003; Lee 2003).

Unfortunately, the grandmother effect cannot explain menopause in porcupines, who assist no one but their own young of the year and who lead a solitary existence. This means there should be no selection against late-acting (postmenopausal) deleterious genes in females, and such genes should accumulate and cause death soon after menopause. Yet this happens neither in humans nor in porcupines.

In 2007, S. D. Tuljapurkar and coworkers proposed an alternate biological explanation for the evolution of menopause. In a paper titled "Why Men Matter," they proposed an explanation based on male–female age differences at marriage. In traditional societies, such as the !Kung and Gambia of Africa, or the Yanomano and Ache of South America, men routinely marry women 5–15 years younger than themselves. They can do this because they can remain fertile as old as age 75, while women experience menopause at age 50–55. Women in such May-September marriages benefit by marrying men who control more resources; men benefit by marrying more-fertile women. But because older men are involved in so many of the matings, they pass on their longevity genes to their offspring, male and female alike. So, indirectly, women owe their postmenopausal survival to their men. But then the plot reverses, and women wind up outliving their men because they are not saddled with testosterone.

Can this scenario explain menopause in porcupines as well? It might. Males don't enter the mating pool until they can battle their way to the top, a process that takes several years. Whereas female porcupines start mating in their second year, males have little success until their fourth year and later (Sweitzer and Holcombe 1993; Sweitzer and Berger 1997). Unfortunately, the relevant data are sketchy. Most crucially, no one knows at this time how long a dominant male maintains his position. In my own Catskill population, male 791 remained in the area for at least 12 years. At his last capture, on August 30, 2003, he weighed a whopping 8.8 kg—the highest weight I have recorded for any porcupine. Yet, even though mating season would be starting in 2 weeks, his testes were not descended. And because male 791 was never radio-collared, there is no information on his lifetime mating success or residence time at the top of the male hierarchy. But the undescended testicles testify that, despite his intimidating weight, he was no longer the top of the pack in 2003.

As Squirrel's life trajectory took her from being a young, inexperienced mother to years of maturity and successful child-rearing to her final years spent alone, without males in mating season or an offspring nursing sweetly in the summer night, some things did not change. Whereas in the human species, grandmothers can pick up and leave for a snow-free life in the sunny south, Squirrel remained constant to her familiar trees, her dens, and her woodland paths.

Paradoxically, Squirrel's physical powers seemed to increase after her menopause. Body condition in porcupines is most easily expressed as a mass-to-length or mass-to-$(\text{length})^2$ ratio (Barthelmess et al. 2006). The weight that animals put on in summer and fall is essential for winter heat production. In the 3 years preceding her menopause, Squirrel's highest annual body weight averaged 6.6 kg. In the 3 years following, the same weight averaged 7.4 kg—an increase that reflects the absence of nursing and pregnancy drains.

Old Age

Up until her final year, Squirrel gave only the most subtle indications of aging. Always a creature of habit, she seemed to spend longer periods of time in favorite feeding trees or rest trees. When we met in the woods, I studied her features for the signs of age I had seen in the old female I had encountered in 1988. But in a meeting in April 2000, her face held a warm, life-observing expression and had the dark color of a healthy adult. Only the faintest signs of white encroached around her nostrils and as rings around her eyes.

Yet not all was as it should be. When I encounter her a month later, I find her on the ground in a thick blackberry stand, not up some resting tree where a porcupine would normally spend the daylight hours (Figure 14.3). As I come closer, Squirrel tries to climb a hop-hornbeam sapling. She gets only 4 feet off the ground when she seems to lose her grip and falls, without apparent injury. She then moves outside the blackberry thicket to a nearby ash tree and starts

Figure 14.3. Squirrel on the ground, fall 2000. (Photo by UR)

climbing again. She moves slowly, feeling for grip in the furrowed bark, and stops often to rest. But she does reach the canopy, where she rests uncomfortably against a sharply angled branch, She is waiting for me to leave, and I do so.

For the rest of her final summer, I find her mostly on the ground. Sometimes she is sheltering in a rock den, where other porcupines would shelter only in winter. Then, at the end of July, her tree-climbing ability seems to return. I am following her radio signal on a sparkling Sunday afternoon, and the signal pulls me uphill toward the upper boundary of our land. This is the heart of Squirrel's territory, where the trees are old and widely spaced, with a rich ground layer of herbs and wildflowers. Here Squirrel has spent her adult life and raised her offspring, nursing them at night, hiding them when they are young and unable to climb, and later watching them ascend the lindens and aspens they love. I reflect that to know a land well, and to be satisfied to live there and nowhere else, is a gift.

But on this July afternoon, Squirrel's signal continues pulling me still higher, to a neighboring forest where lumbering has been in progress. Tree slash obstructs the ground, and the hot sun beats down on brush and blackberry thickets. Then the forest darkens as I approach the Vireo den escarpment. I enter a forest island, a green stretch of woods that has escaped the skidder's chain. Here Squirrel rests in the canopy of a great oak tree that bends gently downhill, and whose thick branches offer comfortable rest.

I search the ground below. Some 30 yards north of Squirrel's oak I find a red trillium, now in fruit. Most of the mass of the trillium is invested not in the leaves and stalk visible aboveground but in a stout rhizome below the ground, which can be as old as the trees above it (Davis 1981; Jules 1988). And to complete the magic, I find nearby two other wildflowers associated with mature forest: a doll's eyes (*Actea alba*) and a dwarf ginseng (*Panax trifolium*). The territory Squirrel occupies is ground she has evaluated and selected. This mountain forest, with old, established trees and herbs has satisfied her for all of her long lifetime.

For the remainder of her final summer and fall, Squirrel moves further downhill, returning to the heart of her territory. During the oak-masting season, she spends several weeks in a giant oak tree, feeding on acorns.

Then the climbing problems recur. She enters a den in early September, missing the year 2000 mating season. On September 30, with fall colors rising to their maximum, I find her back in the sunlight. Rachel and I are following her signal uphill, and almost run into her on the ground. She has been resting at the base of a large sugar maple. A cluster of shiny droppings testifies that she has spent most of the day in this spot.

As I try to approach slowly with my camera, Squirrel moves just as slowly away, quills half-erect. At times, she stops at a tree and grasps the bark in preparation for climbing, but always changes her mind. She is clearly reluctant to climb, and seems to have some trouble walking as well. She continues moving up the hill, in and out of shadow, until I let her disappear from view.

Although I have allowed her to get away, Rachel suggests I might have done more—remove the radiocollar she has worn for most of her life and give her full freedom. I answer that Squirrel's life has become important precisely because she has become the world's oldest monitored porcupine in the wild. Without radio contact, she would slip into the kind of obscurity that hides the lives of most wild animals we share the world with.

Admittedly, Squirrel's biography has enormous gaps. Where was she born? Who were the fathers of her offspring? What were the fates of her offspring, most of whom did not carry radiocollars? Has she survived major traumas like tree falls and predator attacks? Have there been changes in relationship with neighbor porcupines since she underwent menopause? The questions are endless, and answering them requires a much more fine-grained observation schedule than I have been able to maintain.

In October and early November, before first snowfall, Squirrel is in her traditional dens high up the mountain. On November 18, as dusk is falling, I find her some 500 feet downhill from her den site, in a beech grove with a scattering of sugar maple. She is on the ground, clutching the base of a mid-size sugar maple. She starts climbing when I come closer. The tree is an ideal climbing tree, with tightly furrowed bark and with a pronounced slope. Squirrel moves about 10 feet off the ground and then stops, resting on the sloping trunk. She chatters her teeth at me and I depart, leaving 2 Red Delicious apples at the base of her tree.

I return to the grove next day. Squirrel is back in her den, but she has left evidence of last night's work. One of the large apples has been consumed. The striking observation is basal chewing on 12 surrounding beech trees. From most trees, only a small sampling bite has been taken. But two saplings and one adult tree show heavy feeding up to about 16 inches above the ground—the height a porcupine can reach without climbing (Figure 14.4). In no previous winter has Squirrel shown this kind of behavior.

At my next meeting, on December 2, fluffy snow covers the ground. I revisit the beech grove and find feeding scars, all basal, on 16 trees. The remaining apple at the base of the sugar maple is gone, as I confirm by scraping the ground. I find Squirrel in a low-altitude den just above the grove. The den is a poor substitute for her traditional Split Rock den, measuring only some 5 feet deep. In front stand two basal-chewed beech trees. The ground is very steep—marks in the snow indicate that Squirrel had tobogganed downhill.

Figure 14.4. Squirrel's feeding sign at ground level, December 2000. She is no longer able to climb into the canopy. (Photo by UR)

For the rest of December, as snow steadily builds on the mountain, Squirrel occupies various low-altitude dens and basal-chews the bark of surrounding trees. She still has limited ability to climb—I find one beech tree with a feeding zone 6 feet above ground level.

Then, on New Year's day 2001, she has a new address. She is in a hemlock grove at the base of the mountain. While snow elsewhere lies knee-deep, the sheltering canopy of the hemlocks reduces it locally to inches. Squirrel sits inside the hollow trunk of a dead beech tree. Through an opening in the side, I can clearly see her face and radiocollar (Figure 14.5). Her tracks lead to a neighboring hemlock. Porcupines can't feed on hemlock by basal-chewing the bark because it contains high tannin levels. (Hemlock bark was the basis of the nineteenth-century Catskills tanning industry.) To feed there, Squirrel has to climb to feed in the foliage. But live branches can start at just 6 feet above the ground, and they stretch horizontally.

Deep snow covers the mountain on January 27, 2001. The low temperature since my last visit has been 4 °F. I find Squirrel sitting somberly on a low branch of a hemlock just a few feet from her den. A light dusting of snow covers her back. Cut niptwigs in the snow show she has been feeding here.

On March 10, 2001, snow at the cabin measures 22 inches, too deep for the deer, who have all left for the hemlock deeryard at the bottom of the valley. I

Figure 14.5. Squirrel in her hollow-tree den, final view. (Photo by UR)

snowshoe over to the hemlock grove, carrying two apples as gifts. Squirrel is again up a horizontal hemlock branch, this time about 15 feet off the ground. A snow trail leads back to the hollow-beech den. Squirrel, fluffed up on the branch, gives no reaction while I move about, taking photos and planting my apples at the base of her tree. Both apples are gone next morning when I return for my final farewell.

At my next visit, March 31, 2001, she lies buried in the snow, just outside her den. I gently pick up her body and carry it back to the cabin. She gives every indication of health: a high 7.2 kg body weight and a sleek coat with no sign of injury. Later, a careful dissection by Stuart Landry at SUNY-Binghamton shows a stomach full of green foliage—death had not come from starvation.

But it is her skeleton, prepared by technicians at the American Museum of Natural History, that speaks most completely to her condition and to the persecutions of old age. It now becomes obvious why she had so much difficulty climbing during her last year; she was suffering from osteoarthritis in both arms. On the left arm, the heads of both the radius and ulna (the lower arm bones) are severely deformed at the wrist-articulating surfaces. On the right arm, metacarpals (hand bones) 2 and 3 (with vestigial thumb = 1) have fused with three wrist bones (the centrale, the magnum, and the trapezoid). As a result, both arms must have had seriously impaired mobility, forcing her to feed on the ground or in low branches. In confirmation, the front claws are sharp, with little wear.

Another sign of old age shows up in the teeth. Her lower left first molar is worn close to the pulp—the central grinding surfaces have been lost. It will take little further wear in this area to expose the pulp and set up a tooth infection. Why a single tooth has been worn deeper than its neighbors is not obvious—perhaps an earlier injury.

One of her ribs shows a possible healed fracture. Broken ribs are a common consequence of falling from trees. None of her skeletal deformations were life-threatening. Yet, taken together, they must have made her final year full of pain.

In the years that have followed, the distillation of her history becomes perhaps an altered perception of the landscape. I often encounter landmarks associated with Squirrel's life: the massive Split Rock den she spent her winters in, the ledge-trunked linden 766 where she liked to feed and whose root crevices sheltered her babies, the thick old sugar maple that offered a comfortable rest surface and a tree hole to shelter in from rain. I will always see the forest differently.

Squirrel inhabited the forest in all its layers, from the dark forest floor, where seedlings stretch out their small, stiff seed leaves, to the upper branches of the canopy, where light and shadow dance with the wind. It is here that I like to imagine Squirrel resting on an autumn day, half in shadow, half in the sun. Above her, ragged clouds move while she lies perfectly still, swayed gently by her branch, perhaps watching the clouds and the leaves that will soon be falling. When her life ended, I hope she understood how rich it had been, remembering the clouds and the leaves and the changing mountains, with her offspring on distant slopes.

References

Alkon, P.U., and D. Saltz. 1985. Potatoes and nutritional ecology of crested porcupines in a desert biome. J. Appl. Ecol. 22:727–737.

——. 1988. Influence of season and moonlight on temporal activity patterns of Indian crested porcupines (*Hystrix indica*). J. Mammal. 69:71–80.

Allison, M.J., H.M. Cook, and D.A. Stahl. 1987. Characterization of rumen bacteria that degrade dihydroxypyridine compounds produced from mimosine. *In* M. Rose, ed., Herbivore nutrition research, pp. 55–56. Yeerongpilly, Australia: Dept. of Primary Industries, Wool Biology Laboratory.

Anderson, J.F., and L.A. Magnarelli. 1980. Vertebrate host relationships and distribution of Ixodid ticks (Acari: Ixodidae) in Connecticut, USA. J. Med. Entomol. 17:314–323.

Anderson, J.F., L.A. Magnarelli, R.N. Philip, and W. Burgdorfer. 1986. *Rickettsia rickettsii* and *Rickettsia montana* from Ixodid ticks in Connecticut. Am. J. Trop. Med. Hyg. 35:187–191.

Anderson, S., and J.K. Jones, eds. 1967. Recent mammals of the world: A synopsis of families. New York: Ronald Press.

Bakuzis, E.V., and V. Kurmis. 1978. Provisional list of synecological coordinates and selected ecographs of forest and other plant species in Minnesota. Staff paper no. 5. Department of Forest Resources, College of Forestry, University of Minnesota.

Balows, A., and M.W. Jennison. 1949, Thermophilic, cellulose-decomposing bacteria from the porcupine. J. Bacteriol. 57:135.

Barker, I.K., L.R. Lindsay, G.D. Campbell, G.A. Surgeoner, and S.A. McEwen. 1993. The groundhog tick *Ixodes cookei* (Acari: Ixodidae): A poor potential vector of Lyme borreliosis. J. Wildl. Dis. 29(3):416–422.

Barthelmess, E.L., M.L. Phillips, and M.E. Schuckers. 2006. The value of bioelectrical impedance analysis vs. condition indices in predicting body fat stores in North American porcupines (*Erethizon dorsatum*). Can. J. Zool. 84:1712–1720.

Bartlett, C.M., and R.C. Anderson. 1985. The third stage larva of *Molinema arbuta* Highby 1943 (Nematoda) and development of the parasite in the porcupine (*Erethizon dorsatum*). Ann. Parasitol. 60:703–708.

Bauchop, T. 1978. Digestion of leaves in vertebrate arboreal folivores. *In* G.G. Montgomery, ed., The ecology of arboreal folivores, pp. 193–204. Washington, D.C.: Smithsonian Institution Press.

Belovsky, G.E., and P.A. Jordan. 1981. Sodium dynamics and adaptations of a moose population. J. Mammal. 62:613–621.

Berteaux, D., I. Klvana, and C. Trudeau. 2005. Spring to fall weight gain in a northern population of North American porcupines. J. Mammal. 86(3): 514–519.

Besch, D.A. 1987. Animaline. Anim. Kingdom 90(3):51.

Betancourt, J., T. Van Devender, and M. Rose. 1986. Comparison of plant macrofossils in woodrat (*Neotoma* sp.) and porcupine (*Erethizon dorsatum*) middens from the western United States. J. Mammal. 67:266–273.

Bonvicino, C.R., F.C. Almeida, and R. Cerqueira. 2000. The karyotype of *Sphiggurus villosus* (Rodentia: Erethizontidae). Stud. Neotrop. Fauna Environ. 35:81–83.

Bonvicino, C.R., V. Penna-Firme, and E. Braggio. 2002. Molecular and karyotypic evidence of the taxonomic status of *Coendou* and *Sphiggurus* (Rodentia: Hystricognatha). J. Mammal. 83(4):1071–1076.

Bradley, S.R., and D.R. Deavers. 1980. A re-examination of the relationship between thermal conductance and body weight in mammals. Comp. Biochem. Physiol. A 65:465–476.

Brander, R.B. 1973. Life-history notes on the porcupine in a hardwood-hemlock forest in upper Michigan. Mich. Acad. 5:425–433.

Brightsmith, D.J., and R.A. Munoz-Najar. 2004. Avian geophagy and soil characteristics in SE Peru. Biotropica 36:534–543.

Bryant, J.P., J. Tuomi, and P. Niemela. 1988. Environmental constraint of constitutive and long-term inducible defenses in woody plants. *In* K.C. Spencer, ed., Chemical mediation of coevolution, pp. 367–390. San Diego: Academic Press.

Bulstrode, C., J. King, and B. Roper. 1986. What happens to wild animals with broken bones? Lancet 1(8471):29–31.

Burge, B.L. 1966. Vaginal casts passed by captive porcupine. J. Mammal. 47:713–714.

Calder, D.R., and J.S. Bleakney. 1965. Microarthropod ecology of a porcupine-inhabited cave in Nova Scotia. Ecology 46:895–899.

Carey, A.B., W.L. Krinsky, and A.J. Main. 1980. *Ixodes dammini* (Acari: Ixodidae) and associated ticks in south-central Connecticut, USA. J. Med. Entomol. 17:89–99.

Chapman, D.M., and U. Roze. 1997. Functional histology of quill erection in the porcupine (*Erethizon dorsatum*). Can. J. Zool. 75:1–10.

Charles-Dominique, P., M. Atramentowicz, M. Charles-Dominique, H. Gérard, A. Hladik, and M.F. Prévost. 1981. Les mammiferes frugivores arboricoles nocturnes d'une foret Guyanaise: Inter-relations plantes-animaux. Rev. Ecol. 35:341–435.

Clarke, S.H. 1969. Thermoregulatory response of the porcupine, *Erethizon dorsatum,* at low environmental temperatures. Special report, Dept. Forestry and Wildlife Management, University of Massachusetts, Amherst.

Clarke, S.H., and R.B. Brander. 1973. Radiometric determination of porcupine surface temperature. Physiol. Zool. 46:230–237.

Cogan, M.G., and F.C. Rector. 1982. Proximal resorption during metabolic acidosis in the rat. Am. J. Physiol. 242:F499.

Comtois, A., and D. Berteaux. 2005. Impacts of mosquitoes and blackflies on defensive behavior and microhabitat use of the North American porcupine (*Erethizon dorsatum*) in southern Quebec. Can. J. Zool. 83:754–764.

Conn, E.E. 1979. Cyanide and cyanogenic glycosides. *In* G.A. Rosenthal and D.H. Janzen, eds., Herbivores: Their interaction with secondarv plant metabolites, pp. 387–412. Orlando, Fla.: Academic Press.

Conniff, R. 1986. All things considered, a porcupine would rather be left alone. Audubon 88(l):88–93.

Corbet, N.V., and R.J. Van Aarde. 1996. Social organization and space use in the cape porcupine, in a southern African savanna. Afr. J. Ecol. (Nairobi) 34(1):1–14.

Coulter, M.W. 1966. Ecology and management of fishers in Maine. PhD dissertation, State University College of Forestry, Syracuse University.

Craig, E.H., and B.L. Keller. 1986. Movements and home range of porcupines, *Erethizon dorsatum,* in Idaho shrub desert. Can. Field-Nat. 100:167–173.

Crockett, C.M., and J.F. Eisenberg. 1987. Howlers: Variations in group size and demography. *In* B.B. Smuts, D.L. Cheney, R.M. Seyfarth, R.W. Wrangham, and T.T. Struhsaker, eds. Primate societies, pp. 54–68. Chicago: University of Chicago Press.

Curtin, S.H., and D.J. Chivers. 1978. Leaf-eating primates of peninsular Malaysia: The siamang and the dusky leaf-monkey. *In* G.G. Montgomery, ed., The ecology of arboreal folivores, pp. 441–464. Washington, D.C.: Smithsonian Institution Press.

Curtis, J.D. 1941. The silvicultural significance of the porcupine. J. For. 39:583–594.

——. 1944. Appraisal of porcupine damage. J. Wildl. Manage. 8:88–91.

Curtis, J.D., and E L. Kozicky. 1944. Observations on the eastern porcupine. J. Mammal. 25:137–146.

Curtis, J.D., and A.K. Wilson. 1953. Porcupine feeding on ponderosa pine in central Idaho. J. For. 51:339–341.

Dammann, P. 2006. Seneszenz bei Afrikanischen Sandgrabern (Bathyergidae, Rodentia) unter besonderer Berucksichtigung der Gattung *Fukomys.* PhD dissertation, Universitat Duisburg-Essen.

Dathe, H. 1937. Uber den Bau des mannlichen Kopulationsorganes beim Meerschweinschen und anderen hystricomorphen Nagetieren. Gegebaurs Morph. Jahrb. 80:1–65.

——. 1963. Vom Harnspritzen des Ursons (*Erethizon dorsatus*). Z. Saugetierkunde 28:369–375.

Davis, M. 1981. The effect of pollinators, predators, and energy constraints on the floral ecology and evolution of *Trillium erectum.* Oecologia 48:400–406.

DeMatteo, K.E., and H.J. Harlow. 1997. Thermoregulatory responses of the North American porcupine (*Erethizon dorsatum bruneri*) to decreasing ambient temperature and increasing wind speed. Comp. Biochem. Physiol. 116B:339–346.

Denton, D. 1982. The hunger for salt. New York: Springer Verlag.

Denys, N. 1908. The description and natural history of the coasts of N.A. (Acadia). W.F. Ganong, trans. and ed. Publ. Champlain Soc. 2.

de Réaumur, R.A.F. 1727. Observations sur le porc-epic; Extraites de memoires et de lettres de M. Sarrazin, Medecin du Roy a Quebec et Correspondant de L'Academie. Mems. L'Acad. Roy. Sci. (Paris) 1727:383–396.

Devillers, M.S., R.J. Van Aarde, and H.M. Dott. 1994. Habitat utilization by cape porcupines (*Hystrix africaeaustralis*) in a savanna ecosystem. J. Zool. 232(4):539–549.

DeVos, A. 1953. Bobcat preying on porcupine? J. Mammal. 34:129–130.

Diner, B., D. Berteaux, J. Fyles, and R.L. Lindroth. 2009. Behavioral archives link the chemistry and clonal structure of trembling aspen to the food choice of North American porcupine. Oecologia, in press.

Dodds, D.G., A.M. Martell, and R.E. Yescott. 1969. Ecology of the American dog tick, *Dermacentor variabilis* (Say) in Nova Scotia. Can. J. Zool. 47:171–181.

Dodge, W.E. 1967. The biology and life history of the porcupine (*Erethizon dorsatum*) in western Massachusetts. PhD dissertation, University of Massachusetts, Amherst.

——. 1982. Porcupine. *In* J.A. Chapman and G.A. Feldhamer, eds., Wild mammals of North America: Biology, management, and economics, pp. 355–366. Baltimore: Johns Hopkins University Press.

Dodge, W.E., and V.G. Barnes. 1975. Movements, home range and control of porcupines in western Washington. Wildlife leaflet 507. U.S. Department of the Interior, Fish and Wildlife Service, Washington, D.C.

Ducroz, J.-F., V. Volobouev, and L. Granjon. 1998. A molecular perspective on the systematics and evolution of the genus *Arvicanthis* (Rodentia, Muridae): Inferences from complete cyrochrome b sequences. Mol. Phylogenet. Evol. 10(1):104–117.

Duffield, D.A., and R.S. Wells. 1990. Bottlenose dolphins: Comparison of census data from dolphins in captivity. Proc. Eigtheenth Annu. IMATA Conf.

Duffy, D.L., G.E. Bentley, D.L. Drazen, and G.F. Ball. 2000. Effects of testosterone on cell-mediated and humoral immunity in non-breeding adult European starlings. Behav. Ecol. 11(6):654–662.

Duthie, A.G., and J.D. Skinner. 1986. Osteophagia in the Cape porcupine *Hystrix africaeaustralis*. S. Afr. J. Zool. 21:316–318.

Earle, R.D., and K.R. Kramm. 1980. Techniques for age determination in the Canadian porcupine. J. Wildl. Manage. 44:413–419.

——. 1982. Correlation between fisher and porcupine abundance in upper Michigan. Am. Midl. Nat. 107:244–249.

Ebling, F.J., and P.A. Hale. 1970. The control of the mammalian molt. *In* G.K. Benson and J.G. Phillips, eds., Hormones and the environment, pp. 215–238. Cambridge, UK: Cambridge University Press.

Eisenberg, J.F. 1978. The evolution of arboreal folivores in the class Mammalia. *In* G.G. Montgomery, ed., The ecology of arboreal folivores, pp. 135–152. Washington, D.C.: Smithsonian Institution Press.

——. 1983. Behavioral adaptations of higher vertebrates to tropical forests. *In* F.B. Golley, ed., Tropical rain forest ecosystems, pp. 267–278. New York: Elsevier.

Emmons, L.H. 1983. A field study of the African brush-tailed porcupine, *Atherurus africanus,* by radiotelemetry. Mammalia 47:183–194.

Erdoes, R., and A. Ortiz. 1984. American Indian myths and legends. New York: Pantheon Books.

Estrada, A., and R. Coates-Estrada. 1985. A preliminary study of resource overlap between howling monkeys (*Alouatta palliata*) and other arboreal mammals in the tropical rain forest of Los Tuxtlas, Mexico. Am. J. Primatol. 9:27–37.

Evans, F.C. 1985. In praise of natural history. Bull. Ecol. Soc. Amer. 66:455–460.

Fagerstone, K.A., H.P. Tietjen, and O. Williams. 1981. Seasonal variation in the diet of black-tailed prairie dogs. J. Mammal. 60:820–824.

Faulkner, C.E., and W.E. Dodge. 1962. Control of the porcupine in New England. J. For. 60:36–37.

Feeny, P. 1970. Seasonal changes in oak leaf tannins and nutrients as a cause of spring feeding by winter moth caterpillars. Ecology 51:565–581.

——. 1976. Plant apparency and chemical defense. *In* J.W. Wallace and R.L. Mansell, eds., Recent advances in phytochemistry, vol. 10: Biochemical interactions between plants and insects, pp. 1–40. New York: Plenum Press.

Felicetti, L., G.W. Witmer, L.A. Shipley, and C.T. Robbins. 2000. Digestibility, nitrogen excretion, and mean retention time by North American porcupines (*Erethizon dorsatum*) consuming natural forages. Physiol. and Biochem. Zool. 73:772–780.

Fournier, F., and D.W. Thomas. 1999. Thermoregulation and repeatability of oxygen-consumption measurements in winter-acclimatized North American porcupines (*Erethizon dorsatum*). Can. J. Zool. 77:174–202.

Frazier, M.K. 1981. A revision of the fossil Erethizontidae of North America. Bull. Fla. State Mus. Biol. Sci. 27:1–76.

Freeman, R.S. 1949. Notes on the morphology and life cycle of the genus *Monoecocestus* Beddard, 1914 (Cestoda: Anoplocephalidae), from the porcupine. J. Parasitol. 35:605–612.

——. 1952. Biology and life history of *Monoecocestus* Beddard, 1914 (Cestoda: Anoplocephalidae), from the porcupine. J. Parasitol. 38:111–129.

Gabrielson, I.N. 1928. Notes on the habits and behavior of the porcupine in Oregon. J. Mammal. 9:33–38.

Gariepy, E. 1961. Contribution a l'histoire de la medecine Canadienne. Doctorat en Medecine dissertation, Faculte de Medecine de Paris.

Gates, C.C., and R.J. Hudson. 1979. Weight dynamics of free-ranging elk. Agric. For. Bull. 7:80–82.

Gensch, R.H. 1946. Observations on the porcupine in northern Michigan. Unpublished progress report, U.S. Fish and Wildlife Service.

George, W. 1985. Reproductive and chromosomal characters of ctenodactylids as a key to their evolutionary relationships. *In* W.P. Luckett and J.-L. Hartenberger, eds., Evolutionary relationships among rodents, pp. 453–474. New York: Plenum Press.

Gilardi, J.D., S.S. Duffey, C.A. Munn, and L.A. Tell. 1999. Biochemical functions of geophagy in parrots: detoxification of dietary toxins and cytoprotective effects. J. Chem. Ecol. 25:897–922.

Glassman, B.N. 1948. Surface-active agents and their application in bacteriology. Bacteriol. Rev. 12:105–148.

Gosling, L.M. 1980. Reproduction of the Himalayan porcupine *(Hystrix hodgsoni)* in captivity. J. Zool. (Lond.) 192:546–549.

Greenwood, P.J. 1980. Mating systems, philopatry, and dispersal in birds and mammals. Anim. Behav. 28:1140–1162.

Griesemer, S.J., R.M. DeGraaf, and T.K. Fuller. 1994. Effects of excluding porcupines from established winter feeding trees in central Massachusetts. Northeast Wildl. 51:29–33.

Griesemer, S.J., T.K. Fuller, and R.M. DeGraaf. 1996. Denning patterns of porcupines, *Erethizon dorsatum*. Can. Field-Nat. 110(4):634–637.

——. 1998. Habitat use by porcupines *(Erethizon dorsatum)* in central Massachusetts: Effects of topography and forest composition. Am. Midl. Nat. 140:271–279.

Gutterman, Y. 1982. Observations on the feeding habits of the Indian crested porcupine *(Hystrix indica)* and the distribution of some hemicryptophytes and geophytes in the Negev Desert highlands. J. Arid Envir. 5:261–268.

Hale, M.B., and T.K. Fuller. 1996. Porcupine (*Erethizon dorsatum*) demography in central Massachusetts. Can. J. Zool. 74:480–484.

Hamilton, J.B., and G.E. Mestler. 1969. Mortality and survival: Comparison of eunuchs with intact men and women in a mentally retarded population. J. Gerontol. 24:395–411.

Harborne, J.B. 1988. Introduction to ecological biochemistry. 3rd ed. New York: Academic Press.

Harcourt, A.H. 1978. Strategies of emigration and transfer by primates, with particular reference to gorillas. Z. Tierpsych. 48:401–420.

Harder, L.D. 1979. Winter feeding by porcupines in montane forests of southwestern Alberta. Can. Field-Nat. 93:405–410.

Harvey, S., B. Jemiolo, and M. Novotny. 1989. Pattern of volatile compounds in dominant and subordinate male mouse urine. J. Chem. Ecol. 15:2061–2071.

Hawkes, K. 2003. Grandmothers and the evolution of human longevity. Am. J. Hum. Biol. 15:380–400.

Heaney, L.R., and R.W. Thorington. 1978. Ecology of tropical red-tailed squirrel, *Sciurus granatensis,* in the Panama Canal Zone. J. Mammal. 59:846–851.

Hebert, D., and I.M. Cowan. 1971. Natural salt ticks as a part of the ecology of the mountain goat. Can. J. Zool. 49:605–610.

Highby, P.R. 1943a. *Dipetalonema arbuta* n. sp. (Nematoda) from the porcupine, *Erethizon dorsatum* (L). J. Parasitol. 29:239–242.

——. 1943b. Mosquito vectors and larval development of *Dipetalonema arbuta* Highby (Nematoda) from the porcupine, *Erethizon dorsatum.* J. Parasitol. 29:243–252.

Holdo, R.M., J.P. Dudley, and L.R. McDowell. 2002. Geophagy in the African elephant in relation to availability of dietary sodium. J. Mamm. 83:632–664.

Holekamp, K.E. 1986. Proximal causes of natal dispersion in Belding's ground squirrel (*Spermophilus beldingi*). Ecol. Monogr. 56:365–391.

Hoogland, J.L. 1982. Prairie dogs avoid extreme inbreeding. Science 215:1639–1641.

Hooper, E.T. 1961. The glans penis in *Proechimys* and other caviomorph rodents. Occassional papers, Museum of Zoology, University of Michigan, 623:1–18.

Huchon, D., and E.J.P. Douzery. 2001. From the Old World to the New World: A molecular chronicle of the phylogeny and biogeography of Hystricognath rodents. Mol. Phylogenet. Evol. 20:238–251.

Huchon, D., O. Madsen, M.J.J.B. Sibbald, K. Ament, M.J. Stanhope, F. Catzeflis, W.W. de Jong, and E.J.P. Douzery. 2002. Rodent phylogeny and a timescale for the evolution of Glires: Evidence from an extensive taxon sampling using 3 nuclear genes. Mol. Biol. Evol. 19:1053–1065.

Hugot, J.-P. 1982. Sur le genre *Wellcomia* (Oxyuridae, Nematoda), parasite de rongeurs archaiques. Bull. Mus. Nat. Hist. Nat. (Paris) 4:25–48.

——. 2002. New evidence of Hystricognath rodent monophyly from the phylogeny of their pinworms. Bull. Mus. Nat. Hist. Nat. 8:133–138.

Hungate, R.E. 1966. The rumen and its microbes. New York: Academic Press.

Irving, L., and J. Krog. 1954. The body temperature of arctic and subarctic birds and mammals. J. Appl. Physiol. 6:667–680.

Irving, L., H. Krog, and M. Monson. 1955. The metabolism of some Alaskan animals in winter and summer. Physiol. Zool. 28:173–185.

Jellison, W.L. 1933. Parasites of porcupines of the genus *Erethizon* (Rodentia). Trans. Am. Microsc. Soc. 52:42–47.

Jellison, W.L., and K.A. Neiland. 1965. Parasites of Alaskan vertebrates. Norman: University of Oklahoma Research Institution.

Jemiolo, B., S. Harvey, and M. Novotny. 1986. Promotion of the Whitten effect in female mice by synthetic analogs of male urinary constituents. Proc. Nat. Acad. Sci. 83:4576–4579.

Johns, T., and M. Duquette. 1991. Traditional detoxification of acorn bred with clay. Ecol. Food Nutr. 25:221–228.

Johnson, J.L., and R.H. McBee. 1967. The porcupine caecal fermentation. J. Nutr. 91:540–546.

Jones, R.J., and R.G. Megarrity. 1986. Successful transfer of DHP-degrading bacteria from Hawaiian goats to Australian ruminants to overcome the toxicity of *Leucaena.* Aust. Vet. J. 63:259–262.

Jordan, P.A., D.B. Botkin, A.S. Dominski, H.S. Lowendorf, and G.E. Belovsky. 1973. Sodium as a critical nutrient of the moose of Isle Royale. Proc. N. Am. Moose Conf. Workshop 9:13–42.

Jules, E.S. 1988. Habitat fragmentation and demographic change for a common plant: *Trillium* in old-growth forest. Ecology 79(5):1645–1656.

Kirkwood, T.B.L. 1977. Evolution of ageing. Nature 270:301–304.

Kirkwood, T.B.L., and S.N. Austad. 2000. Why do we age? Nature 408:233–237.

Koenigswald, W. von, 1985. Evolutionary trends in the enamel of rodent incisors. *In* L.P. Luckett and J.-L. Hartenberger, eds., Evolutionarv relationships among rodents, pp. 403–422. New York: Plenum Press.

Krefting, L.W., J.H. Stoeckeler, B.J. Bradle, and W.D. Fitzwater. 1962. Porcupine-timber relationships in the lake states. J. For. 60:325–330.

Landry, S. Jr. 1957. The interrelationships of the New and Old World hystricomorph rodents. University of California Publications in Zoology 56:1–118.

——. 1977. Book review: The biology of hystricomorph rodents, I.W. Rowlands and Barbara J. Weir. J. Mammal. 58:459–461.

Lawson Handley, L.J., and N. Perrin. 2007. Advances in our understanding of mammalian sex-biased dispersal. Mol. Ecol. 16:1559–1578.

Lee, R.D. 2003. Rethinking the evolutionary theory of aging: Transfers, not births, shape senescence in social species. Proc. Natl. Acad. Sci. U.S.A. 100:9637–9642.

Li, G., U. Roze, and D.C. Locke. 1997. Warning odor of the North American porcupine (*Erethizon dorsatum*). J. Chem. Ecol. 23:2737–2754.

Lindholm, J.S., J.M. McCormick, S.W. Colton, and D.T. Downing. 1981. Variation of skin surface lipid composition among mammals. Comp. Biochem. Physiol. 69B:75–78.

Luckett, W.P., and J.-L. Hartenberger. 1985. Comments and conclusions. *In* W.P. Luckett and J.-L. Hartenberger, eds., Evolutionarv relationships among rodents, pp. 685–712. New York: Plenum Press.

Lyford, C.A. 1943. Ojibwa crafts (Chippewa). Indian handicraft pamphlet no. 5. Branch of Education, Bureau of Indian Affairs, U.S. Department of the Interior, Washington, D.C.

Mabille, G., D. Berteaux, D.W. Thomas, and D. Fortin. In press. Behavioural responses of porcupines to the heterogeneity of the thermal environment during the boreal winter.

Mabille, G., S. Descamps, and D. Berteaux. In press. Predation as a possible mechanism relating winter weather to population dynamics in a North American porcupine population.

Macdonald, D. 1984. The encyclopedia of mammals. New York: Facts on File.

MacDonald, D.R. 1952. Some observations on porcupine attacks on jack pine. The Annual Ring, Faculty of Forestry, University of Toronto, pp. 17–18.

Magnarelli, L.A., J.F. Anderson, W. Burgdorfer, R.N. Philip, and W.A. Chappell. 1985. Spotted fever group rickettsiae in in mature and adult ticks (Acari: Ixodidae) from a focus of Rocky Mountain spotted fever in Connecticut. Can. J. Microbiol. 31:1131–1135.

Main, A.J., A.B. Carey, M.G. Carey, and R.H. Goodwin. 1982. Immature *Ixodes dammini* (Acari: Ixodidae) on small animals in Connecticut, USA. J. Med. Entomol. 19:655–664.

Marshall, W.H. 1951. Accidental death of a porcupine. J. Mammal. 32:221.

Maser, C., and R.S. Rohweder. 1983. Winter food habits of cougars from northeastern Oregon. Great Basin Nat. 43:425–428.

Mattson, W.J. Jr. 1980. Herbivory in relation to plant nitrogen content. Ann. Rev. Ecol. Syst. 11:119–161.

McDade, H.C., and W.B. Crandell. 1958. Perforation of the gastrointestinal tract by an unusual body—a porcupine quill. New Engl. J. Med. 258:746–747.

McEvoy, J.S. 1982. Comparative myology of the pectoral and pelvic appendages of the North Amdrican porcupine (*Erethizon dorsatum*) and the prehensile-tailed porcupine (*Coendou prehensilis*). Bull. Amer. Mus. Nat. Hist. 173:337–421.

Mead-Briggs, A.R., and R.J.C. Page. 1975. Records of anoplocephaline cestodes from wild rabbits and hares collected throughout Great Britain. J. Helminthol. 49:49–56.

Mech, L.D. 1987. Age, season, distance, direction, and social aspects of wolf dispersal from a Minnesota pack. D.D. Chepko-Sade and Z.T. Halpin, eds., pp. 55–74. Chicago and London: University of Chicago Press.

Medawar, P.B. 1952. An unsolved problem of biology. London: Lewis Publishers.

Merriam, C.H. 1884. The Mammals of the Adirondack Region. New York: L.S. Foster Press.

Miniano, A.M., M.P. Hreon, S.L. Murphy, and K.D. Kochanek. 2007. Final data for 2004. National Vital Statistics Reports v. 55, no. 19. Atlanta: Center for Disease Control and Prevention.

Mirand, E.A., and A.R. Shadle. 1953. Gross anatomy of the male reproductive system of the porcupine. J. Mamm. 34:210–220.

Mohr, Erna. 1965. Altweltliche Stachelschweine. Wittenberg Lutherstadt: A. Ziemsen Verlag.

Monctti, L., A. Massolo, A. Sforzi, and S. Lovari. 2005. Site selection and fidelity by crested porcupines for denning. Ethol. Ecol. Evol. 17:149–159.

Montgomery, G.G., and Y.D. Lubin. 1978. Movements of *Coendou prehensilis* in the Venezuelan llanos. J. Mammal. 59:887–888.

Montgomery, G.G., and M.E. Sunquist. 1978. Habitat selection and use by two-toed and three-toed sloths. G. Gene Montgomery, ed., pp. 329–359. The Ecology of Arborel Folivores. Washington D.C.: Smithsonian Institution Press.

Morin, P., D. Berteaux, and I. Klvana. 2005. Hierarchical habitat selection by North American porcupines in a southern boreal forest. Can. J. Zool. 83:1333–1342.

Morris, P. 1983. Hedgehogs. The Oil Mills, Weybridge, Surrey: Whittet Books.

Morse, M. 1903. Synopsis of North American invertebrates 29: the Trichodectidae. Amer. Nat. 37:618.

Murie, O.J. 1926. The porcupine in northern Alaska. J. Mamm. 7:109–113.

Murphy, W.J., E. Elzirik, W.E. Johnson, Y.P. Zhang, O.A. Ryder, and S.J. O'Brien. 2001. Molecular phylogenetics and the origins of placental mammals. Nature 409:612–618.

Nicolaides, N. 1974. Skin lipids: their biochemical uniqueness. Science 186:19–26.

Novotny, M.V. 2003. Pheromones, binding proteins, and receptor responses in rodents. Biochem. Soc. Trans. 31:117–123.

Odle, R. 1971. Porcupine quillwork in the Great Lakes area. The Explorer 13:23–28.

Oliveira, P.A. 2006. The ecology of female thin-spined porcupines *Chaetomys subspinosus* (Olfers 1818) (Rodentia: Erethizontidae) in restinga forests at the Parque Estadual Paulo César Vinha, southeast Brazil. M.S. thesis, Pontifica Universidate Catolica de Minas Gerais.

Oliver, J.H., M.R. Owsley, H.J. Hutcheson, A.M. James, C. Chen, W.S. Irby, E.M.Dotson, and D.K. McLain. 1993. Conspecificity of the ticks *Ixodes scapularis* and *I. dammini* (Acari: Ixodidae.) J. Med. Entomol. 30(1):54–63.

Olsen, O.W., and C.D. Tollman. 1951. *Wellcomia evaginata* Smitt 1908 (Oxyuridae: Nematoda) of porcupines in mule deer, *Odocoileus hemionus*, in Colorado. Proc. Helm. Soc. Wash. 18:120–122.

Orchard, W.C. 1971. The technique of porcupine-quill decoration among the North American Indians. 2nd ed. Contributions from the Museum of the American Indian, no. 4. Heye Foundation, New York.

Owens, I.P.F. 2002. Sex differences in mortality rate. Science 297:2008–2009.

Packer, C., M. Tatar, and A. Collins. 1998. Reproductive cessation in female mammals. Nature 392:807–811.

Payette, S. 1987. Recent porcupine expansion at tree line: A dendroecological analysis. Can. J. Zool. 65:551–557.

Payne, D.D., and D.C. O'Meara. 1958. *Sarcoptes scabiei* infestation of a porcupine. J. Wildl. Manage. 22:321–322.

Po-Chedley, D.S., and A.R. Shadle. 1955. Pelage of the porcupine, *Erethizon dorsatum dorsatum*. J. Mammal. 36:84–95.

Pocock, R.L. 1922. On the external characteristics of some hystricomorph rodents. Proc. Zool. Soc. London 1922:365–427.

Powell, R.A. 1978. A comparison of fisher and weasel hunting behavior. Carnivore 1:28–34.

——. 1993. The fisher: Life history, ecology, and behavior, 2nd ed., Minneapolis: University of Minnesota Press.

Powell, R.A., and R.B. Brander. 1977. Adaptations of fishers and porcupines to their predator-prey system. *In* R.L. Phillips and C. Jonkel, eds., Proc. 1975 Predator Symp., pp. 45–53.

Poux, C., P. Chevret, D. Huchon, W.W. de Jong, and E.P. Douzery. 2006. Arrival and diversification of Caviomorph rodents and Platyrrhine primates in South America. J. Syst. Biol. 55(2):228–244.

Prescott, J.H. 1959. Rafting of jack rabbit on kelp. J. Mammal. 40:443–444.

Pusey, A.E., and C. Packer. 1987. Dispersal and philopatry. *In* B.B. Smuts, D.L. Cheney, R.M. Seyfarth, R.W. Wrangham, and T.T. Struhsaker, eds. Primate societies, pp. 250–266. Chicago: University of Chicago Press.

Raeburn, P. 1987. Point of no return? Not so. Cincinnati Enquirer, Jan. 23.

Rahm, U. 1962. L'elevage et la reproduction en captivite de l'*Atherurus africanus* (Rongeurs: Hystricidae). Mammalia 29:1–9.

Reeks, W.A. 1942. Notes on the Canada porcupine in the Maritime provinces. For. Chron. 18:182–187.

Reynolds, H.G. 1957. Porcupine behavior in the desert-shrub type of Arizona. J. Mammal. 33:418–419.

Richter, C.P., and B. Barelare. 1938. Nutritional requirements of pregnant and lactating rats studied by the self-selection method. Endocrinology 23:15–24.

Robbins, C.T. 1983. Wildlife feeding and nutrition. New York: Academic Press.

Roberts, M., S. Brand, and E. Maliniak. 1985. The biology of captive prehensile-tailed porcupines, *Coendou prehensilis*. J. Mammal. 66:476–482.

Rogers, P. 2000. The old men of the sea. San Jose Mercury News, Dec. 19.

Rose, B.D. 1984. Clinical physiology of acid-base and electrolyte disorders. 2nd ed. New York: McGraw-Hill.

Rousseau, J. 1957. Michel Sarrazin, Jean-Francois Gaulthier et l'etude prelinneenne de la Flore Canadienne. *In* Les Botanistes Francais en Amerique du Nord avant 1850, ed., pp. 149–155. Paris: Centre National de la Recherche Scientifique.

Rowlands, I.W., and B.J. Weir. 1974. The biology of hystricomorph rodents. *In* Symp. Zool. Soc. London 34.
Roze, U. 1984. Winter foraging by individual porcupines. Can. J. Zool. 62:2425–2428.
——. 1985. How to select, climb, and eat a tree. Nat. Hist. 94(5):63–68.
——. 1987. Denning and winter range of the porcupine. Can. J. Zool. 65:981–986.
——. 2002. A facilitated release mechanism for quills of the North American porcupine (*Erethizon dorsatum*). J. Mammal. 83(2):381–385.
——. 2004. Risk factors for injury in porcupines. Wildl. Rehabil. 21:61–63.
——. 2006. Smart weapons. Nat. Hist. 115(2):48–53.
Roze, U., K.T. Leung, E. Nix, G. Burton, and D.M. Chapman. In press. Microanatomy and bacterial flora of the perineal glands of the North American porcupine, *Erethizon dorsatum* (L.). Can. J. Zool.
Roze, U., D.C. Locke, and N. Vatakis. 1990. Antibiotic properties of porcupine quills. J. Chem. Ecol. 16:725–734.
Ryder, J.W., R.R. Pinger, and T. Glancy. 1992. Inability of *Ixodes cookei* and *Amblyomma americanum* nymphs (Acari: Ixodidae) to transmit *Borrelia burgdorferi*. J. Med. Entomol. 29(3):525–530.
Sackett, L.W. 1913. The Canada porcupine: A study of the learning process. Behav. Monogr. 2:1–84.
Sauer, G.C. 1980. Manual of skin diseases. 4th ed. Philadelphia: J.B. Lippincott.
Seastedt, T.R., and D.A. Crossley Jr. 1981. Sodium dynamics in forest ecosystems and the animal starvation hypothesis. Am. Nat. 117:1029–1034.
Seton, E.T. 1929. Lives of game animals, Vol. 4. Boston: Doubleday, Page.
Shadle, A.R. 1946. Copulation in the porcupine. J. Wildl. Manage. 10:159–162.
——. 1948. Gestation period in the porcupine, *Erethizon dorsatum dorsatum*. J. Mammal. 29:162–164.
——. 1950. The North American porcupine up to date. Ward's Nat. Sci. Bull. 24:5–11.
——. 1951. Laboratory copulations and gestations of porcupine, *Erethizon dorsatum*. J. Mammal. 32:219–221.
——. 1952. Sexual maturity and first recorded copulation of a 16-month male porcupine, *Erethizon dorsatum dorsatum*. J. Mammal. 33:239–241.
Shadle, A.R., M. Smelzer, and M. Metz. 1946. The sex reactions of porcupines (*Erethizon d. dorsatum*) before and after copulation. J. Mammal. 27:116–121.
Shanley, D.P., and T.B.L. Kirkwood. 2001. Evolution of the human menopause. BioEssays 23:283–287.
Shapiro, J. 1949. Ecological and life history notes on the porcupine in the Adirondacks. J. Mammal. 30:247–257.
Sherris, J.C. 1984a. Normal microbial flora. *In* J.C. Sherris, ed., Medical microbiology, pp. 50–58. New York: Elsevier.
——. 1984b. Skin and wound infections. *In* J.C. Sherris, ed., Medical microbiology, pp. 555–561. New York: Elsevier.
Simpson, G.G. 1980. Splendid isolation: The curious history of South American mammals. New Haven: Yale University Press.
Smith, G.W. 1982. Habitat use by porcupines in a ponderosa pine–Douglas fir forest in northeastern Oregon. Northwest Sci. 56:236–240.
Smithers, R.H.N. 1983. The mammals of the southern African subregion. Pretoria: University of Pretoria Press.

Speck, F.G., and R.W. Dexter. 1951. Utilization of animals and plants by the Micmac Indians of New Brunswick. J. Wash. Acad. Sci. 41:250–259.

Speer, R.J., and T.G. Dilworth. 1978. Porcupine winter foods and utilization in central New Brunswick. Can. Field-Nat. 92:271–274.

Spencer, D.A. 1949. Porcupine problems of the Nicolet National Forest. Unpublished special report, U.S. Fish and Wildlife Service.

Stricklan, D., J.T. Flinders, and R.G. Cates. 1995. Factors affecting selection of winter food and roosting resources by porcupines in Utah. Great Basin Nat. 55(1):29–36.

Struhsaker, T.T. 1975. The red Colobus monkey. Chicago: University of Chicago Press.

Struhsaker, T.T., and L. Leland. 1987. Colobines infanticide by adult males. *In* B.B. Smuts, D.L. Cheney, R.M. Seyfarth, R.W. Wrangham, and T.T. Struhsaker, eds., Primate societies, pp. 83–97. Chicago: University of Chicago Press.

Sun, L., D. Muller-Schwarze, and B.A. Schulte. 2000. Dispersal pattern and effective population size of the beaver. Can. J. Zool. 78:393–398.

Sutton, J.F. 1972. Notes on skeletal variation, tooth replacement, and cranial suture closure of the porcupine (*Erethizon dorsatum*). Tulane Stud. Zool. Bot. 17:56–62.

Sweitzer, R.A. 2003. Breeding movements and reproductive activities of porcupines in the Great Basin desert. West. N. Am. Nat. 63:1–10.

Sweitzer, R.A., and J. Berger. 1993. Seasonal dynamics of mass and body condition in Great Basin porcupines (*Erethizon dorsatum*). J. Mammal. 74(1):198–203.

——. 1997. Sexual dimorphism and evidence for intrasexual selection from quill impalements, injuries, and mate guarding in porcupine (*Erethizon dorsatum*). Can. J. Zool. 75:847–854.

——. 1998. Evidence for female-biased dispersal in North American porcupines (*Erethizon dorsatum*). J. Zool. (Lond.) 244:159–166.

Sweitzer, R.A., and D.W. Holcombe. 1993. Serum-progesterone levels and pregnancy rates in Great Basin porcupines (*Erethizon dorsatum*). J. Mamm. 74:769–776.

Sweitzer, R.A., S.H. Jenkins, and Joel Berger. 1997. Near-extinction of porcupines by mountain lions and consequences of ecosystem change in the Great Basin desert. Conserv. Biol. 11(6):1407–1417.

Taylor, W.P. 1935. Ecology and life history of the porcupine (*Erethizon epixanthum*) as related to the forests of Arizona and the southwestern United States. Univ. Ariz. Bull. 6:1–177.

Teale, E.W. 1965. Wandering through winter. New York: Dodd, Mead and Co.

Tenneson, C., and L.W. Oring. 1985. Winter food preferences of porcupines. J. Wildl. Manage. 49:28–33.

Terborgh, J., and C.H. Janson. 1986. The socioecology of primate groups. Ann. Rev. Ecol. Syst. 17:111–135.

Thornton, R.F., P.R. Bird, M. Somers, and R.J. Moir. 1970. Urea excretion in ruminants, III: The role of the hind-gut (caecum and colon). Aust. J. Agric. Res. 21:345–352.

Thwaites, R.G., ed. 1907. Jesuit relations and allied documents: Travels and explorations of the Jesuit missionaries in New France 1610–1791. Cleveland: Burrows Bros.

Tindal, J.S., and G.S. Knaggs. 1970. Environmental stimuli and the mammary gland. *In* G.K. Benson and J.G. Phillips, eds., Hormones and the environment, pp. 239–258. Cambridge, UK: Cambridge University Press.

Tohme, H., and G. Tohme. 1981. Quelques donnees anatomiques sur le porc-epic *Hystrix indica* Kerr 1792 (Rodentia). Mammalia 45:363–371.

Tryon, C.A. Jr. 1947. Behavior and postnatal development of a porcupine. J. Wildl. Manage. 11:282–283.

Tuljapurkar, S.D, C.O. Puleston, and M.D. Gurven. 2007. Why men matter: Mating patterns drive evolution of human lifespan. PLoS ONE 2(8):e785 [online].

Van Aarde, R.J. 1985. Reproduction in captive female Cape porcupines (*Hystrix africaeaustralis*) J. Reprod. Fertil. 75:577–582.

——. 1987. Reproduction in the cape porcupine, *Hystrix africaeaustralis:* An ecological perspective. S. Afr. J. Sci. 83(10):605–607.

Van Aarde, R.J., and J.D. Skinner. 1986. Reproductive biology of the male Cape porcupine, *Hystrix africaeaustratis.* J. Reprod. Fertil. 76:545–552.

Van Aarde, R.J., and V. Van Wyk. 1991. Reproductive inhibition in the cape porcupine, *Hystrix africaeaustralis.* J. Reprod. Fertil. 92:13–19.

Van Jaarsveld, A.S., and A. Knight-Eloff. 1984. Digestion in the porcupine *Hystrix africaeaustralis.* S. Afr. J. Zool. 19:109–112.

Vaughan, T.A. 1978. Mammalogy. 2nd ed. Philadelphia: Saunders College.

Vilela, R.V., T. Machado, K. Ventura, V. Fagundes, M.J. de J. Silva, and Y. Yonenaga-Yassuda. 2009. The taxonomic status of the endangered thin-spined porcupine, *Chaetomys* subspinosus (Olfers, 1818), based on molecular and karyological data. BMC Evolutionary Biology 9(29):1–17.

Vispo, C., and I.D. Hume. 1995. The digestive tract and digestive function in the North American porcupine and beaver. Can. J. Zool. 73:967–974.

Vom Saal, F.S., C.E. Finch, and J.F. Nelson. 1994. Natural history and mechanisms of reproductive aging in humans, laboratory rodents, and other selected vertebrates. E. Knobil and J.D. Neill, eds., pp. 1213–1314. The Physiology of Reproduction, 2nd ed. New York: Raven Press.

Voss, R. 1979. Male accessory glands and the evolution of copulatory plugs in rodents. Occasional Papers, Museum of Zoology, University of Michigan, 689:1–26.

Voss, R.S., and R. Angermann. 1997. Revisionary notes on neotropical porcupines (Rodentia: Erethizontidae), 1: Type material described by Olfers (1818) and Kuhl (1820) in the Berlin Zoological Museum. Am. Mus. Novit. 3214:1–42.

Wallace, K., and P. Henry. 1985. Return of a Catskills native. Conservationist 40(3):16–19.

Waser, P.M., and W.T. Jones. 1983. Natal philopatry among solitary mammals. Q. Rev. Biol. 58:355–390.

Weeks, H.P. Jr., and C.M. Kirkpatrick. 1976. Adaptations of white-tailed deer to naturally occurring sodium deficiencies. J. Wild. Manage. 40:610–625.

——. 1978. Salt preferences and sodium drive phenology in fox squirrels and woodchucks. J. Mammal. 59:531–542.

Weir, B.J. 1974. Reproductive characteristics of hystricomorph rodents. Symp. Zool. Soc. Lond. 34:265–301.

Whitten, W.K. 1956. Modification of the estrous cycle of the mouse by external stimuli associated with the male. J. Endocrinol. 13:399–404.

Wiles, G.J., and H.P. Weeks Jr. 1986. Movements and use patterns of white-tailed deer visiting natural licks. J. Wildl. Manage. 50:487–496.

Williams, G.C. 1957. Pleiotropy, natural selection, and the evolution of senescence. Evolution 11:398–411.

Wilson, D.E., and D.M. Reeder. 2005. Mammal species of the world. 3rd ed. Baltimore: Johns Hopkins University Press.

Wood, J.W., P.E. Smouse, and J.C. Long. 1985. Sex-specific dispersal patterns in two human populations of highland New Guinea. Am. Nat. 125:747–768.

Woods, C.A. 1972. Comparative myology of jaw, hyoid, and pectoral appendicular regions of New World and Old World hystricomorph rodents. Bull. Am. Mus. Nat. Hist. 147:117–198.

——. 1973. *Erethizon dorsatum*. Mamm. Spec. 29:1–6.

Wrangham, R.W. 1986. Ecology and social relationships in two species of chimpanzee. *In* D.I. Rubenstein and R.W. Wrangham, eds., Ecological aspects of social evolution, pp. 352–378. Princeton: Princeton University Press.

Wyss, O., B.J. Ludwig, and R.B. Joiner. 1945. The fungistatic and fungicidal action of fatty acids and related compounds. Arch. Biochem. 7:415–425.

Zuk, M., and K.A. McKean. 1996. Sex differences in parasite infections: Patterns and processes. J. Parasitol. 26(10):1009–1023.

Index